Practical Linear Algebra
for Data Science
From Core Concepts to Applications
Using Python

Mike X Cohen

Beijing · Boston · Farnham · Sebastopol · Tokyo

Practical Linear Algebra for Data Science
by Mike X Cohen

Published by O'Reilly Media, Inc., 1005 Gravenstein Highway North, Sebastopol, CA 95472.

O'Reilly books may be purchased for educational, business, or sales promotional use. Online editions are also available for most titles (*http://oreilly.com*). For more information, contact our corporate/institutional sales department: 800-998-9938 or *corporate@oreilly.com*.

Acquisitions Editor: Jessica Haberman	**Indexer:** Ellen Troutman
Development Editor: Shira Evans	**Interior Designer:** David Futato
Production Editor: Jonathon Owen	**Cover Designer:** Karen Montgomery
Copyeditor: Piper Editorial Consulting, LLC	**Illustrator:** Kate Dullea
Proofreader: Shannon Turlington	

September 2022:   First Edition

Revision History for the First Edition
2022-09-01:   First Release
2023-07-14:   Second Release
2023-09-01:   Third Release

See *https://www.oreilly.com/catalog/errata.csp?isbn=0636920641025* for release details.

978-1-098-12061-0

[LSI]

Table of Contents

Preface

Conventions Used in This Book

The following typographical conventions are used in this book:

Italic
: Indicates new terms, URLs, email addresses, filenames, and file extensions.

`Constant width`
: Used for program listings, as well as within paragraphs to refer to program elements such as variable or function names, databases, data types, environment variables, statements, and keywords.

> This element signifies a general note.

> This element indicates a warning or caution.

Using Code Examples

Supplemental material (code examples, exercises, etc.) is available for download at *https://github.com/mikexcohen/LinAlg4DataScience*.

If you have a technical question or a problem using the code examples, please send email to *bookquestions@oreilly.com*.

This book is here to help you get your job done. In general, if example code is offered with this book, you may use it in your programs and documentation. You do not need to contact us for permission unless you're reproducing a significant portion of the code. For example, writing a program that uses several chunks of code from this book does not require permission. Selling or distributing examples from O'Reilly books does require permission. Answering a question by citing this book and quoting example code does not require permission. Incorporating a significant amount of example code from this book into your product's documentation does require permission.

We appreciate, but generally do not require, attribution. An attribution usually includes the title, author, publisher, and ISBN. For example: "*Practical Linear Algebra for Data Science* by Mike X. Cohen (O'Reilly). Copyright 2022 Syncxpress BV, 978-1-098-12061-0."

If you feel your use of code examples falls outside fair use or the permission given above, feel free to contact us at *permissions@oreilly.com*.

O'Reilly Online Learning

 For more than 40 years, *O'Reilly Media* has provided technology and business training, knowledge, and insight to help companies succeed.

Our unique network of experts and innovators share their knowledge and expertise through books, articles, and our online learning platform. O'Reilly's online learning platform gives you on-demand access to live training courses, in-depth learning paths, interactive coding environments, and a vast collection of text and video from O'Reilly and 200+ other publishers. For more information, visit *https://oreilly.com*.

How to Contact Us

Please address comments and questions concerning this book to the publisher:

O'Reilly Media, Inc.
1005 Gravenstein Highway North
Sebastopol, CA 95472
800-998-9938 (in the United States or Canada)
707-829-0515 (international or local)
707-829-0104 (fax)

We have a web page for this book, where we list errata, examples, and any additional information. You can access this page at *https://oreil.ly/practical-linear-algebra*.

Email *bookquestions@oreilly.com* with comments or technical questions about this book.

For news and information about our books and courses, visit *https://oreilly.com*.

Find us on LinkedIn: *https://linkedin.com/company/oreilly-media*

Follow us on Twitter: *https://twitter.com/oreillymedia*

Watch us on YouTube: *https://youtube.com/oreillymedia*

Acknowledgments

I have a confession: I really dislike writing acknolwedgments sections. It's not because I lack gratitude or believe that I have no one to thank—quite the opposite: I have too many people to thank, and I don't know where to begin, who to list by name, and who to leave out. Shall I thank my parents for their role in shaping me into the kind of person who wrote this book? Perhaps their parents for shaping my parents? I remember my fourth-grade teacher telling me I should be a writer when I grow up. (I don't remember her name and I'm not sure when I will grow up, but perhaps she had some influence on this book.) I wrote most of this book during remote-working trips to the Canary Islands; perhaps I should thank the pilots who flew me there? Or the electricians who installed the wiring at the coworking spaces? Perhaps I should be grateful to Özdemir Pasha for his role in popularizing coffee, which both facilitated and distracted me from writing. And let's not forget the farmers who grew the delicious food that sustained me and kept me happy.

You can see where this is going: my fingers did the typing, but it took the entirety and history of human civilization to create me and the environment that allowed me to write this book—and that allowed you to read this book. So, thanks humanity!

But OK, I can also devote one paragraph to a more traditional acknowledgments section. Most importantly, I am grateful to all my students in my live-taught university and summer-school courses, and my Udemy online courses, for trusting me with their education and for motivating me to continue improving my explanations of applied math and other technical topics. I am also grateful for Jess Haberman, the acquisitions editor at O'Reilly who made "first contact" to ask if I was interested in writing this book. Shira Evans (development editor), Jonathon Owen (production editor), Elizabeth Oliver (copy editor), Kristen Brown (manager of content services), and two expert technical reviewers were directly instrumental in transforming my keystrokes into the book you're now reading. I'm sure this list is incomplete because other people who helped publish this book are unknown to me or because I've forgotten them due to memory loss at my extreme old age.[1] To anyone reading this who feels they made even an infinitesimal contribution to this book: thank you.

---

1 LOL, I was 42 when I wrote this book.

## CHAPTER 1

Introduction

What Is Linear Algebra and Why Learn It?

Linear algebra has an interesting history in mathematics, dating back to the 17th century in the West and much earlier in China. Matrices—the spreadsheets of numbers at the heart of linear algebra—were used to provide a compact notation for storing sets of numbers like geometric coordinates (this was Descartes's original use of matrices) and systems of equations (pioneered by Gauss). In the 20th century, matrices and vectors were used for multivariate mathematics including calculus, differential equations, physics, and economics.

But most people didn't need to care about matrices until fairly recently. Here's the thing: computers are extremely efficient at working with matrices. And so, modern computing gave rise to modern linear algebra. Modern linear algebra is computational, whereas traditional linear algebra is abstract. Modern linear algebra is best learned through code and applications in graphics, statistics, data science, AI, and numerical simulations, whereas traditional linear algebra is learned through proofs and pondering infinite-dimensional vector spaces. Modern linear algebra provides the structural beams that support nearly every algorithm implemented on computers, whereas traditional linear algebra is often intellectual fodder for advanced mathematics university students.

Welcome to modern linear algebra.

Should you learn linear algebra? That depends on whether you want to understand algorithms and procedures, or simply apply methods that others have developed. I don't mean to disparage the latter—there is nothing intrinsically wrong with using tools you don't understand (I am writing this on a laptop that I can use but could not build from scratch). But given that you are reading a book with this title in the O'Reilly book collection, I guess you either (1) want to know how algorithms work or

(2) want to develop or adapt computational methods. So yes, you should learn linear algebra, and you should learn the modern version of it.

About This Book

The purpose of this book is to teach you modern linear algebra. But this is not about memorizing some key equations and slugging through abstract proofs; the purpose is to teach you how to *think* about matrices, vectors, and operations acting upon them. You will develop a geometric intuition for why linear algebra is the way it is. And you will understand how to implement linear algebra concepts in Python code, with a focus on applications in machine learning and data science.

Many traditional linear algebra textbooks avoid numerical examples in the interest of generalizations, expect you to derive difficult proofs on your own, and teach myriad concepts that have little or no relevance to application or implementation in computers. I do not write these as criticisms—abstract linear algebra is beautiful and elegant. But if your goal is to use linear algebra (and mathematics more generally) as a tool for understanding data, statistics, deep learning, image processing, etc., then traditional linear algebra textbooks may seem like a frustrating waste of time that leave you confused and concerned about your potential in a technical field.

This book is written with self-studying learners in mind. Perhaps you have a degree in math, engineering, or physics, but need to learn how to implement linear algebra in code. Or perhaps you didn't study math at university and now realize the importance of linear algebra for your studies or work. Either way, this book is a self-contained resource; it is not solely a supplement for a lecture-based course (though it could be used for that purpose).

If you were nodding your head in agreement while reading the past three paragraphs, then this book is definitely for you.

If you would like to take a deeper dive into linear algebra, with more proofs and explorations, then there are several excellent texts that you can consider, including my own *Linear Algebra: Theory, Intuition, Code* (Sincxpress BV).[1]

Prerequisites

I have tried to write this book for enthusiastic learners with minimal formal background. That said, nothing is ever learned truly from from scratch.

---

1 Apologies for the shameless self-promotion; I promise that's the only time in this book I'll subject you to such an indulgence.

Math

You need to be comfortable with high-school math. Just basic algebra and geometry; nothing fancy.

Absolutely zero calculus is required for this book (though differential calculus is important for applications where linear algebra is often used, such as deep learning and optimization).

But most importantly, you need to be comfortable thinking about math, looking at equations and graphics, and embracing the intellectual challenge that comes with studying math.

Attitude

Linear algebra is a branch of mathematics, ergo this is a mathematics book. Learning math, especially as an adult, requires some patience, dedication, and an assertive attitude. Get a cup of coffee, take a deep breath, put your phone in a different room, and dive in.

There will be a voice in the back of your head telling you that you are too old or too stupid to learn advanced mathematics. Sometimes that voice is louder and sometimes softer, but it's always there. And it's not just you—everyone has it. You cannot suppress or destroy that voice; don't even bother trying. Just accept that a bit of insecurity and self-doubt is part of being human. Each time that voice speaks up is a challenge for you to prove it wrong.

Coding

This book is focused on linear algbera applications in code. I wrote this book for Python, because Python is currently the most widely used language in data science, machine learning, and related fields. If you prefer other languages like MATLAB, R, C, or Julia, then I hope you find it straightforward to translate the Python code.

I've tried to make the Python code as simple as possible, while still being relevant for applications. Chapter 16 provides a basic introduction to Python programming. Should you go through that chapter? That depends on your level of Python skills:

Intermediate/advanced (>1 year coding experience)
   Skip Chapter 16 entirely, or perhaps skim it to get a sense of the kind of code that will appear in the rest of the book.

Some knowledge (<1 year experience)
   Please work through the chapter in case there is material that is new or that you need to refresh. But you should be able to get through it rather briskly.

Go through the chapter in detail. Please understand that this book is not a complete Python tutorial, so if you find yourself struggling with the code in the content chapters, you might want to put this book down, work through a dedicated Python course or book, then come back to this book.

Mathematical Proofs Versus Intuition from Coding

The purpose of studying math is, well, to understand math. How do you understand math? Let us count the ways:

Rigorous proofs

A proof in mathematics is a sequence of statements showing that a set of assumptions leads to a logical conclusion. Proofs are unquestionably important in pure mathematics.

Visualizations and examples

Clearly written explanations, diagrams, and numerical examples help you gain intuition for concepts and operations in linear algebra. Most examples are done in 2D or 3D for easy visualization, but the principles also apply to higher dimensions.

The difference between these is that formal mathematical proofs provide rigor but rarely intuition, whereas visualizations and examples provide lasting intuition through hands-on experience but can risk inaccuracies based on specific examples that do not generalize.

Proofs of important claims are included, but I focus more on building intuition through explanations, visualizations, and code examples.

And this brings me to mathematical intuition from coding (what I sometimes call "soft proofs"). Here's the idea: you assume that Python (and libraries such as NumPy and SciPy) correctly implements the low-level number crunching, while you focus on the principles by exploring many numerical examples in code.

A quick example: we will "soft-prove" the commutivity principle of multiplication, which states that $a \times b = b \times a$:

```
a = np.random.randn()
b = np.random.randn()
a*b - b*a
```

This code generates two random numbers and tests the hypothesis that swapping the order of multiplication has no impact on the result. The third line of would print out `0.0` if the commutivity principle is true. If you run this code multiple times and always get `0.0`, then you have gained intuition for commutivity by seeing the same result in many different numerical examples.

To be clear: intuition from code is no substitute for a rigorous mathematical proof. The point is that "soft proofs" allow you to understand mathematical concepts without having to worry about the details of abstract mathematical syntax and arguments. This is particularly advantageous to coders who lack an advanced mathematics background.

The bottom line is that *you can learn a lot of math with a bit of coding.*

Code, Printed in the Book and Downloadable Online

You can read this book without looking at code or solving code exercises. That's fine, and you will certainly learn something. But don't be disappointed if your knowledge is superficial and fleeting. If you really want to *understand* linear algebra, you need to solve problems. That's why this book comes with code demonstrations and exercises for each mathematical concept.

Important code is printed directly in the book. I want you to read the text and equations, look at the graphs, and *see the code* at the same time. That will allow you to link concepts and equations to code.

But printing code in a book can take up a lot of space, and hand-copying code on your computer is tedious. Therefore, only the key code lines are printed in the book pages; the online code contains additional code, comments, graphics embellishments, and so on. The online code also contains solutions to the coding exercises (all of them, not only the odd-numbered problems!). You should definitely download the code and go through it while working through the book.

All the code can be obtained from the GitHub site *https://github.com/mikexcohen /LinAlg4DataScience*. You can clone this repository or simply download the entire repository as a ZIP file (you do not need to register, log in, or pay to download the code).

I wrote the code using Jupyter notebook in Google's Colab environment. I chose to use Jupyter because it's a friendly and easy-to-use environment. That said, I encourage you to use whichever Python IDE you prefer. The online code is also provided as raw *.py* files for convenience.

Code Exercises

Math is not a spectator sport. Most math books have countless paper-and-pencil problems to work through (and let's be honest: no one does all of them). But this book is all about *applied* linear algebra, and no one applies linear algebra on paper! Instead, you apply linear algebra in code. Therefore, in lieu of hand-worked problems and tedious proofs "left as an exercise to the reader" (as math textbook authors love to write), this book has lots of code exercises.

The code exercises vary in difficulty. If you are new to Python and to linear algebra, you might find some exercises really challenging. If you get stuck, here's a suggestion: have a quick glance at my solution for inspiration, then put it away so you can't see my code, and continue working on your own code.

When comparing your solution to mine, keep in mind that there are many ways to solve problems in Python. Ariving at the correct answer is important; the steps you take to get there are often a matter of personal coding style.

How to Use This Book (for Teachers and Self Learners)

There are three environments in which this book is useful:

Self-learner

I have tried to make this book accessible to readers who want to learn linear algebra on their own, outside a formal classroom environment. No additional resources or online lectures are necessary, although of course there are myriad other books, websites, YouTube videos, and online courses that students might find helpful.

Primary textbook in a data science class

This book can be used as a primary textbook in a course on the math underlying data science, machine learning, AI, and related topics. There are 14 content chapters (excluding this introduction and the Python appendix), and students could be expected to work though one to two chapters per week. Because students have access to the solutions to all exercises, instructors may wish to supplement the book exercises with additional problem sets.

Secondary textbook in a math-focused linear algebra course

This book could also be used as a supplement in a mathematics course with a strong focus on proofs. In this case, the lectures would focus on theory and rigorous proofs while this book could be referenced for translating the concepts into code with an eye towards applications in data science and machine learning. As I wrote above, instructors may wish to provide supplementary exercises because the solutions to all the book exercises are available online.

## CHAPTER 2

Vectors, Part 1

Vectors provide the foundations upon which all of linear algebra (and therefore, the rest of this book) is built.

By the end of this chapter, you will know all about vectors: what they are, what they do, how to interpret them, and how to create and work with them in Python. You will understand the most important operations acting on vectors, including vector algebra and the dot product. Finally, you will learn about vector decompositions, which is one of the main goals of linear algebra.

Creating and Visualizing Vectors in NumPy

In linear algebra, a *vector* is an ordered list of numbers. (In abstract linear algebra, vectors may contain other mathematical objects including functions; however, because this book is focused on applications, we will only consider vectors comprising numbers.)

Vectors have several important characteristics. The first two we will start with are:

Dimensionality
    The number of numbers in the vector

Orientation
    Whether the vector is in *column orientation* (standing up tall) or *row orientation* (laying flat and wide)

Dimensionality is often indicated using a fancy-looking $\mathbb{R}^N$, where the $\mathbb{R}$ indicates real-valued numbers (cf. $\mathbb{C}$ for complex-valued numbers) and the N indicates the dimensionality. For example, a vector with two elements is said to be a member of $\mathbb{R}^2$. That special $\mathbb{R}$ character is made using latex code, but you can also write R², R2, or R^2.

Equation 2-1 shows a few examples of vectors; please determine their dimensionality and orientation before reading the subsequent paragraph.

Equation 2-1. Examples of column vectors and row vectors

$$\mathbf{x} = \begin{bmatrix} 1 \\ 4 \\ 5 \\ 6 \end{bmatrix}, \quad \mathbf{y} = \begin{bmatrix} .3 \\ -7 \end{bmatrix}, \quad \mathbf{z} = \begin{bmatrix} 1 & 4 & 5 & 6 \end{bmatrix}$$

Here are the answers: $\mathbf{x}$ is a 4D column vector, $\mathbf{y}$ is a 2D column vector, and $\mathbf{z}$ is a 4D row vector. You can also write, e.g., $\mathbf{x} \in \mathbb{R}^4$, where the $\in$ symbol means "is contained in the set of."

Are $\mathbf{x}$ and $\mathbf{z}$ the same vector? Technically they are different, even though they have the same elements in the same order. See "Does Vector Orientation Matter?" on page 9 for more discussion.

You will learn, in this book and throughout your adventures integrating math and coding, that there are differences between math "on the chalkboard" versus implemented in code. Some discrepancies are minor and inconsequential, while others cause confusion and errors. Let me now introduce you to a terminological difference between math and coding.

I wrote earlier that the *dimensionality* of a vector is the number of elements in that vector. However, in Python, the dimensionality of a vector or matrix is the number of geometric dimensions used to print out a numerical object. For example, all of the vectors shown above are considered "two-dimensional arrays" in Python, regardless of the number of elements contained in the vectors (which is the mathematical dimensionality). A list of numbers without a particular orientation is considered a 1D array in Python, regardless of the number of elements (that array will be printed out as a row, but, as you'll see later, it is treated differently from row vectors). The mathematical dimensionality—the number of elements in the vector—is called the *length* or the *shape* of the vector in Python.

This inconsistent and sometimes conflicting terminology can be confusing. Indeed, terminology is often a sticky issue at the intersection of different disciplines (in this case, mathematics and computer science). But don't worry, you'll get the hang of it with some experience.

When referring to vectors, it is common to use lowercase bolded Roman letters, like $\mathbf{v}$ for "vector v." Some texts use italics (v) or print an arrow on top ($\vec{v}$).

Linear algebra convention is to assume that vectors are in column orientation unless otherwise specified. Row vectors are written as $\mathbf{w}^{\mathrm{T}}$. The T indicates the *transpose operation*, which you'll learn more about later; for now, suffice it to say that the transpose operation transforms a column vector into a row vector.

Does Vector Orientation Matter?

Do you really need to worry about whether vectors are column- or row-oriented, or orientationless 1D arrays? Sometimes yes, sometimes no. When using vectors to store data, orientation usually doesn't matter. But some operations in Python can give errors or unexpected results if the orientation is wrong. Therefore, vector orientation is important to understand, because spending 30 minutes debugging code only to realize that a row vector needs to be a column vector is guaranteed to give you a headache.

Vectors in Python can be represented using several data types. The list type may seem like the simplest way to represent a vector—and it is for for some applications. But many linear algebra operations won't work on Python lists. Therefore, most of the time it's best to create vectors as NumPy arrays. The following code shows four ways of creating a vector:

```
asList  = [1,2,3]
asArray = np.array([1,2,3]) # 1D array
rowVec  = np.array([ [1,2,3] ]) # row
colVec  = np.array([ [1],[2],[3] ]) # column
```

The variable asArray is an *orientationless* array, meaning it is neither a row nor a column vector but simply a 1D list of numbers in NumPy. Orientation in NumPy is given by brackets: the outermost brackets group all of the numbers together into one object. Then, each additional set of brackets indicates a row: a row vector (variable rowVec) has all numbers in one row, while a column vector (variable colVec) has multiple rows, with each row containing one number.

We can explore these orientations by examining the shapes of the variables (inspecting variable shapes is often very useful while coding):

```
print(f'asList:  {np.shape(asList)}')
print(f'asArray: {asArray.shape}')
print(f'rowVec:  {rowVec.shape}')
print(f'colVec:  {colVec.shape}')
```

Here's what the output looks like:

```
asList:  (3,)
asArray: (3,)
rowVec:  (1, 3)
colVec:  (3, 1)
```

The output shows that the 1D array `asArray` is of size (3), whereas the orientation-endowed vectors are 2D arrays and are stored as size (1,3) or (3,1) depending on the orientation. Dimensions are always listed as (rows,columns).

Geometry of Vectors

Ordered list of numbers is the algebraic interpretation of a vector; the geometric interpretation of a vector is a straight line with a specific length (also called *magnitude*) and direction (also called *angle*; it is computed relative to the positive *x*-axis). The two points of a vector are called the tail (where it starts) and the head (where it ends); the head often has an arrow tip to disambiguate from the tail.

You may think that a vector encodes a geometric coordinate, but vectors and coordinates are actually different things. They are, however, concordant when the vector starts at the origin. This is called the *standard position* and is illustrated in Figure 2-1.

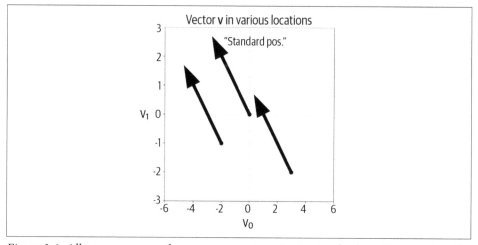

Figure 2-1. All arrows express the same vector. A vector in standard position has its tail at the origin and its head at the concordant geometric coordinate.

Conceptualizing vectors either geometrically or algebraically facilitates intuition in different applications, but these are simply two sides of the same coin. For example, the geometric interpretation of a vector is useful in physics and engineering (e.g., representing physical forces), and the algebraic interpretation of a vector is useful in data science (e.g., storing sales data over time). Oftentimes, linear algebra concepts are learned geometrically in 2D graphs, and then are expanded to higher dimensions using algebra.

Operations on Vectors

Vectors are like nouns; they are the characters in our linear algebra story. The fun in linear algebra comes from the verbs—the actions that breathe life into the characters. Those actions are called *operations*.

Some linear algebra operations are simple and intuitive and work exactly how you'd expect (e.g., addition), whereas others are more involved and require entire chapters to explain (e.g., singular value decomposition). Let's begin with simple operations.

Adding Two Vectors

To add two vectors, simply add each corresponding element. Equation 2-2 shows an example:

Equation 2-2. Adding two vectors

$$\begin{bmatrix} 4 \\ 5 \\ 6 \end{bmatrix} + \begin{bmatrix} 10 \\ 20 \\ 30 \end{bmatrix} = \begin{bmatrix} 14 \\ 25 \\ 36 \end{bmatrix}$$

As you might have guessed, vector addition is defined only for two vectors that have the same dimensionality; it is not possible to add, e.g., a vector in $\mathbb{R}^3$ with a vector in $\mathbb{R}^5$.

Vector subtraction is also what you'd expect: subtract the two vectors element-wise. Equation 2-3 shows an example:

Equation 2-3. Subtracting two vectors

$$\begin{bmatrix} 4 \\ 5 \\ 6 \end{bmatrix} - \begin{bmatrix} 10 \\ 20 \\ 30 \end{bmatrix} = \begin{bmatrix} -6 \\ -15 \\ -24 \end{bmatrix}$$

Adding vectors is straightforward in Python:

```
v = np.array([4,5,6])
w = np.array([10,20,30])
u = np.array([0,3,6,9])
vPlusW = v+w
uPlusW = u+w # error! dimensions mismatched!
```

Does vector orientation matter for addition? Consider Equation 2-4:

Equation 2-4. Can you add a row vector to a column vector?

$$\begin{bmatrix} 4 \\ 5 \\ 6 \end{bmatrix} + \begin{bmatrix} 10 & 20 & 30 \end{bmatrix} = ?$$

You might think that there is no difference between this example and the one shown earlier—after all, both vectors have three elements. Let's see what Python does:

```
v = np.array([[4,5,6]]) # row vector
w = np.array([[10,20,30]]).T # column vector
v+w

>> array([[14, 15, 16],
          [24, 25, 26],
          [34, 35, 36]])
```

The result may seem confusing and inconsistent with the definition of vector addition given earlier. In fact, Python is implementing an operation called *broadcasting*. You will learn more about broadcasting later in this chapter, but I encourage you to spend a moment pondering the result and thinking about how it arose from adding a row and a column vector. Regardless, this example shows that orientation is indeed important: *two vectors can be added together only if they have the same dimensionality **and** the same orientation.*

Geometry of Vector Addition and Subtraction

To add two vectors geometrically, place the vectors such that the tail of one vector is at the head of the other vector. The summed vector traverses from the tail of the first vector to the head of the second (graph A in Figure 2-2). You can extend this procedure to sum any number of vectors: simply stack all the vectors tail-to-head, and then the sum is the line that goes from the first tail to the final head.

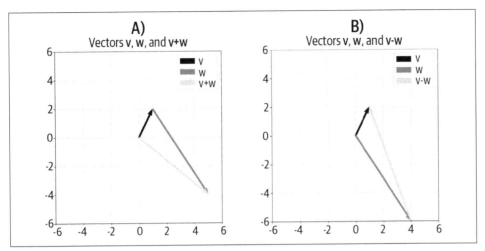

Figure 2-2. *The sum and difference of two vectors*

Subtracting vectors geometrically is slightly different but equally straightforward: line up the two vectors such that their tails are at the same coordinate (this is easily accomplished by having both vectors in standard position); the difference vector is the line that goes from the head of the "negative" vector to the head of the "positive" vector (graph B in Figure 2-2).

Do not underestimate the importance of the geometry of vector subtraction: it is the basis for orthogonal vector decomposition, which in turn is the basis for linear least squares, which is one of the most important applications of linear algebra in science and engineering.

Vector-Scalar Multiplication

A *scalar* in linear algebra is a number on its own, not embedded in a vector or matrix. Scalars are typically indicated using lowercase Greek letters such as α or λ. Therefore, vector-scalar multiplication is indicated as, for example, $\beta\mathbf{u}$.

Vector-scalar multiplication is very simple: multiply each vector element by the scalar. One numerical example (Equation 2-5) will suffice for understanding:

Equation 2-5. Vector-scalar multiplication (or: scalar-vector multiplication)

$$\lambda = 4, \quad \mathbf{w} = \begin{bmatrix} 9 \\ 4 \\ 1 \end{bmatrix}, \quad \lambda\mathbf{w} = \begin{bmatrix} 36 \\ 16 \\ 4 \end{bmatrix}$$

I wrote earlier that the data type of a variable storing a vector is sometimes important and sometimes unimportant. Vector-scalar multiplication is an example where data type matters:

```
s = 2
a = [3,4,5] # as list
b = np.array(a) # as np array
print(a*s)
print(b*s)

>> [ 3, 4, 5, 3, 4, 5 ]
>> [ 6 8 10 ]
```

The code creates a scalar (variable s) and a vector as a list (variable a), then converts that into a NumPy array (variable b). The asterisk is overloaded in Python, meaning its behavior depends on the variable type: scalar multiplying a list repeats the list s times (in this case, twice), which is definitely *not* the linear algebra operation of scalar-vector multiplication. When the vector is stored as a NumPy array, however, the asterisk is interpreted as element-wise multiplication. (Here's a small exercise for you: what happens if you set s = 2.0, and why?[1]) Both of these operations (list repetition and vector-scalar multiplication) are used in real-world coding, so be mindful of the distinction.

Scalar-Vector Addition

Adding a scalar to a vector is not formally defined in linear algebra: they are two separate kinds of mathematical objects and cannot be combined. However, numerical processing programs like Python will allow adding scalars to vectors, and the operation is comparable to scalar-vector multiplication: the scalar is added to each vector element. The following code illustrates the idea:

```
s = 2
v = np.array([3,6])
s+v
>> [5 8]
```

---

[1] a*s throws an error, because list repetition can only be done using integers; it's not possible to repeat a list 2.72 times!

The geometry of vector-scalar multiplication

Why are scalars called "scalars"? That comes from the geometric interpretation. Scalars scale vectors without changing their direction. There are four effects of vector-scalar multiplication that depend on whether the scalar is greater than 1, between 0 and 1, exactly 0, or negative. Figure 2-3 illustrates the concept.

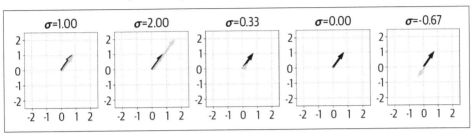

Figure 2-3. The same vector (black arrow) multiplied by different scalars σ (gray line; shifted slightly for visibility)

I wrote earlier that scalars do not change the direction of the vector. But the figure shows that the vector direction flips when the scalar is negative (that is, its angle rotates by 180°. That might seem a contradiction, but there is an interpretation of vectors as pointing along an infinitely long line that passes through the origin and goes to infinity in both directions (in the next chapter, I'll call this a "one-dimensional subspace"). In that sense, the "rotated" vector still points along the same infinite line and thus the negative scalar does not change the direction. This interpretation is important for matrix spaces, eigenvectors, and singular vectors, all of which are introduced in later chapters.

Vector-scalar multiplication in combination with vector addition leads directly to *vector averaging*. Averaging vectors is the same as averaging numbers: sum and divide by the number of numbers. So, to average two vectors, add them and then scalar multiply by .5. In general, to average N vectors, sum them and scalar multiply the result by *1/N*.

Transpose

You already learned about the transpose operation: it converts column vectors into row vectors, and vice versa. Let me provide a slightly more formal definition that will generalize to transposing matrices (a topic in Chapter 5).

A matrix has rows and columns; therefore, each matrix element has a (*row,column*) index. The transpose operation simply swaps those indices. This is formalized in Equation 2-6:

Equation 2-6. The transpose operation

$$\mathbf{m}^T_{i,j} = \mathbf{m}_{j,i}$$

Vectors have either one row or one column, depending on their orientation. For example, a 6D row vector has $i = 1$ and j indices from 1 to 6, whereas a 6D column vector has i indices from 1 to 6 and $j = 1$. So swapping the i,j indices swaps the rows and columns.

Here's an important rule: transposing twice returns the vector to its original orientation. In other words, $\mathbf{v}^{TT} = \mathbf{v}$. That may seem obvious and trivial, but it is the keystone of several important proofs in data science and machine learning, including creating symmetric covariance matrices as the data matrix times its transpose (which in turn is the reason why a principal components analysis is an orthogonal rotation of the data space…don't worry, that sentence will make sense later in the book!).

Vector Broadcasting in Python

Broadcasting is an operation that exists only in modern computer-based linear algebra; this is not a procedure you would find in a traditional linear algebra textbook.

Broadcasting essentially means to repeat an operation multiple times between one vector and each element of another vector. Consider the following series of equations:

$$[1\ 1] + [10\ 20]$$
$$[2\ 2] + [10\ 20]$$
$$[3\ 3] + [10\ 20]$$

Notice the patterns in the vectors. We can implement this set of equations compactly by condensing those patterns into vectors [1 2 3] and [10 20], and then broadcasting the addition. Here's how it looks in Python:

```
v = np.array([[1,2,3]]).T # col vector
w = np.array([[10,20]])   # row vector
v + w # addition with broadcasting

>> array([[11, 21],
          [12, 22],
          [13, 23]])
```

Here again you can see the importance of orientation in linear algebra operations: try running the code above, changing v into a row vector and w into a column vector.[2]

Because broadcasting allows for efficient and compact computations, it is used often in numerical coding. You'll see several examples of broadcasting in this book, including in the section on *k*-means clustering (Chapter 4).

Vector Magnitude and Unit Vectors

The *magnitude* of a vector—also called the *geometric length* or the *norm*—is the distance from tail to head of a vector, and is computed using the standard Euclidean distance formula: the square root of the sum of squared vector elements (see Equation 2-7). Vector magnitude is indicated using double-vertical bars around the vector: $\| \mathbf{v} \|$.

Equation 2-7. The norm of a vector

$$\| \mathbf{v} \| = \sqrt{\sum_{i=1}^{n} v_i^2}$$

Some applications use squared magnitudes (written $\| \mathbf{v} \|^2$), in which case the square root term on the right-hand side drops out.

Before showing the Python code, let me explain some more terminological discrepancies between "chalkboard" linear algebra and Python linear algebra. In mathematics, the dimensionality of a vector is the number of elements in that vector, while the length is a geometric distance; in Python, the function len() (where len is short for *length*) returns the *dimensionality* of an array, while the function np.norm() returns the geometric length (magnitude). In this book, I will use the term *magnitude* (or *geometric length*) instead of *length* to avoid confusion:

```
v = np.array([1,2,3,7,8,9])
v_dim = len(v)  # math dimensionality
v_mag = np.linalg.norm(v) # math magnitude, length, or norm
```

There are some applications where we want a vector that has a geometric length of one, which is called a *unit vector*. Example applications include orthogonal matrices, rotation matrices, eigenvectors, and singular vectors.

A unit vector is defined as $\| \mathbf{v} \| = 1$.

Needless to say, lots of vectors are not unit vectors. (I'm tempted to write "most vectors are not unit vectors," but there is an infinite number of unit vectors and nonunit vectors, although the set of infinite nonunit vectors is larger than the set of

---

2 Python still broadcasts, but the result is a 3 × 2 matrix instead of a 2 × 3 matrix.

infinite unit vectors.) Fortunately, any nonunit vector has an associated unit vector. That means that we can create a unit vector in the same direction as a nonunit vector. Creating an associated unit vector is easy; you simply scalar multiply by the reciprocal of the vector norm (Equation 2-8):

Equation 2-8. Creating a unit vector

$$\widehat{\mathbf{v}} = \frac{1}{\| \mathbf{v} \|} \mathbf{v}$$

You can see the common convention for indicating unit vectors ($\widehat{\mathbf{v}}$) in the same direction as their parent vector **v**. Figure 2-4 illustrates these cases.

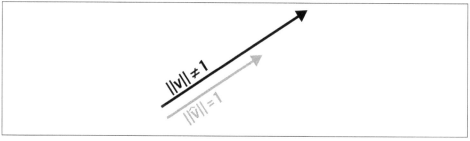

Figure 2-4. A unit vector (gray arrow) can be crafted from a nonunit vector (black arrow); both vectors have the same angle but different magnitudes

Actually, the claim that "*any* nonunit vector has an associated unit vector" is not entirely true. There is a vector that has nonunit length and yet has no associated unit vector. Can you guess which vector it is?[3]

I'm not showing Python code to create unit vectors here, because that's one of the exercises at the end of this chapter.

The Vector Dot Product

The *dot product* (also sometimes called the *inner product*) is one of the most important operations in all of linear algebra. It is the basic computational building block from which many operations and algorithms are built, including convolution, correlation, the Fourier transform, matrix multiplication, linear feature extraction, signal filtering, and so on.

---

3 The zeros vector has a length of 0 but no associated unit vector, because it has no direction and because it is impossible to scale the zeros vector to have nonzero length.

There are several ways to indicate the dot product between two vectors. I will mostly use the common notation $\mathbf{a}^\mathrm{T}\mathbf{b}$ for reasons that will become clear after learning about matrix multiplication. In other contexts you might see $\mathbf{a} \cdot \mathbf{b}$ or $\langle \mathbf{a}, \mathbf{b} \rangle$.

The dot product is a single number that provides information about the relationship between two vectors. Let's first focus on the algorithm to compute the dot product, and then I'll discuss how to interpret it.

To compute the dot product, you multiply the corresponding elements of the two vectors, and then sum over all the individual products. In other words: element-wise multiplication and sum. In Equation 2-9, $\mathbf{a}$ and $\mathbf{b}$ are vectors, and a_i indicates the ith element of $\mathbf{a}$.

Equation 2-9. Dot product formula

$$\delta = \sum_{i=1}^{n} a_i b_i$$

You can tell from the formula that the dot product is valid only between two vectors of the same dimensionality. Equation 2-10 shows a numerical example:

Equation 2-10. Example dot product calculation

$$[1 \ 2 \ 3 \ 4] \cdot [5 \ 6 \ 7 \ 8] = 1 \times 5 + 2 \times 6 + 3 \times 7 + 4 \times 8$$
$$= 5 + 12 + 21 + 32$$
$$= 70$$

Irritations of Indexing

Standard mathematical notation, and some math-oriented numerical processing programs like MATLAB and Julia, start indexing at 1 and stop at N, whereas some programming languages like Python and Java start indexing at 0 and stop at $N - 1$. We need not debate the merits and limitations of each convention—though I do sometimes wonder how many bugs this inconsistency has introduced into human civilization—but it is important to be mindful of this difference when translating formulas into Python code.

There are multiple ways to implement the dot product in Python; the most straightforward way is to the use the `np.dot()` function:

```
v = np.array([1,2,3,4])
w = np.array([5,6,7,8])
np.dot(v,w)
```

Note About np.dot()

The function `np.dot()` does not actually implement the vector dot product; it implements matrix multiplication, which is a collection of dot products. This will make more sense after learning about the rules and mechanisms of matrix multiplication (Chapter 5). If you want to explore this now, you can modify the previous code to endow both vectors with orientations (row versus column). You will discover that the output is the dot product only when the first input is a row vector and the second input is a column vector.

Here is an interesting property of the dot product: scalar multiplying one vector scales the dot product by the same amount. We can explore this by expanding the previous code:

```
s = 10
np.dot(s*v,w)
```

The dot product of v and w is 70, and the dot product using `s*v` (which, in math notation, would be written as $\sigma\mathbf{v}^T\mathbf{w}$) is 700. Now try it with a negative scalar, e.g., `s = -1`. You'll see that the dot product magnitude is preserved but the sign is reversed. Of course, when `s = 0` then the dot product is zero.

Now you know how to compute the dot product. What does the dot product mean and how do we interpret it?

The dot product can be interpreted as a measure of *similarity* or *mapping* between two vectors. Imagine that you collected height and weight data from 20 people, and you stored those data in two vectors. You would certainly expect those variables to be related to each other (taller people tend to weigh more), and therefore you could expect the dot product between those two vectors to be large. On the other hand, the magnitude of the dot product depends on the scale of the data, which means the dot product between data measured in grams and centimeters would be larger than the dot product between data measured in pounds and feet. This arbitrary scaling, however, can be eliminated with a normalization factor. In fact, the normalized dot product between two variables is called the *Pearson correlation coefficient*, and it is one of the most important analyses in data science. More on this in Chapter 4!

The Dot Product Is Distributive

The distributive property of mathematics is that $a(b + c) = ab + ac$. Translated into vectors and the vector dot product, it means that:

$$\mathbf{a}^T(\mathbf{b} + \mathbf{c}) = \mathbf{a}^T\mathbf{b} + \mathbf{a}^T\mathbf{c}$$

In words, you would say that the dot product of a vector sum equals the sum of the vector dot products.

The following Python code illustrates the distributivity property:

```
a = np.array([ 0,1,2 ])
b = np.array([ 3,5,8 ])
c = np.array([ 13,21,34 ])

# the dot product is distributive
res1 = np.dot( a, b+c )
res2 = np.dot( a,b ) + np.dot( a,c )
```

The two outcomes res1 and res2 are the same (with these vectors, the answer is 110), which illustrates the distributivity of the dot product. Notice how the mathematical formula is translated into Python code; translating formulas into code is an important skill in math-oriented coding.

Geometry of the Dot Product

There is also a geometric definition of the dot product, which is the product of the magnitudes of the two vectors, scaled by the cosine of the angle between them (Equation 2-11).

Equation 2-11. Geometric definition of the vector dot product

$$\alpha = \cos\left(\theta_{\mathbf{v},\mathbf{w}}\right) \|\mathbf{v}\| \|\mathbf{w}\|$$

Equation 2-9 and Equation 2-11 are mathematically equivalent but expressed in different forms. The proof of their equivalence is an interesting exercise in mathematical analysis, but would take about a page of text and relies on first proving other principles including the Law of Cosines. That proof is not relevant for this book and so is omitted.

Notice that vector magnitudes are strictly positive quantities (except for the zeros vector, which has $\|\mathbf{0}\| = 0$), while the cosine of an angle can range between -1 and $+1$. This means that the sign of the dot product is determined entirely by the geometric relationship between the two vectors. Figure 2-5 shows five cases of the dot product sign, depending on the angle between the two vectors (in 2D for visualization, but the principle holds for higher dimensions).

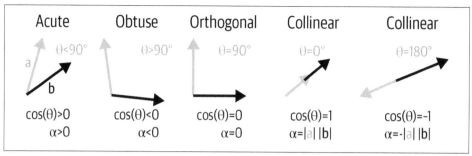

Figure 2-5. *The sign of the dot product between two vectors reveals the geometric relationship between those vectors*

Memorize This: Orthogonal Vectors Have a Zero Dot Product

Some math teachers insist that you shouldn't memorize formulas and terms, and instead should understand procedures and proofs. But let's be honest: memorization is an important and inescapable part of learning mathematics. Fortunately, linear algebra isn't excessively memorization-heavy, but there are a few things you'll simply need to commit to memory.

Here is one: orthogonal vectors have a dot product of zero (that claim goes both ways—when the dot product is zero, then the two vectors are orthogonal). So, the following statements are equivalent: two vectors are orthogonal; two vectors have a dot product of zero; two vectors meet at a 90° angle. Repeat that equivalence until it's permanently etched into your brain.

Other Vector Multiplications

The dot product is perhaps the most important, and most frequently used, way to multiply vectors. But there are several other ways to multiply vectors.

Hadamard Multiplication

This is just a fancy term for element-wise multiplication. To implement Hadamard multiplication, each corresponding element in the two vectors is multiplied. The product is a vector of the same dimensionality as the two multiplicands. For example:

$$\begin{bmatrix} 5 \\ 4 \\ 8 \\ 2 \end{bmatrix} \odot \begin{bmatrix} 1 \\ 0 \\ .5 \\ -1 \end{bmatrix} = \begin{bmatrix} 5 \\ 0 \\ 4 \\ -2 \end{bmatrix}$$

In Python, the asterisk indicates element-wise multiplication for two vectors or matrices:

```
a = np.array([5,4,8,2])
b = np.array([1,0,.5])
a*b
```

Try running that code in Python and...uh oh! Python will give an error. Find and fix the bug. What have you learned about Hadamard multiplication from that error? Check the footnote for the answer.[4]

Hadamard multiplication is a convenient way to organize multiple scalar multiplications. For example, imagine you have data on the number of widgets sold in different shops and the price per widget at each shop. You could represent each variable as a vector, and then Hadamard-multiply those vectors to compute the widget revenue *per shop* (this is different from the total revenue across *all shops*, which would be computed as the dot product).

Outer Product

The outer product is a way to create a matrix from a column vector and a row vector. Each row in the outer product matrix is the row vector scalar multiplied by the corresponding element in the column vector. We could also say that each column in the product matrix is the column vector scalar multiplied by the corresponding element in the row vector. In Chapter 6, I'll call this a "rank-1 matrix," but don't worry about the term for now; instead, focus on the pattern illustrated in the following example:

$$\begin{bmatrix} a \\ b \\ c \end{bmatrix} \begin{bmatrix} d & e \end{bmatrix} = \begin{bmatrix} ad & ae \\ bd & be \\ cd & ce \end{bmatrix}$$

Using Letters in Linear Algebra

In middle school algebra, you learned that using letters as abstract placeholders for numbers allows you to understand math at a deeper level than arithmetic. Same concept in linear algebra: teachers sometimes use letters inside matrices in place of numbers when that facilitates comprehension. You can think of the letters as variables.

---

4 The error is that the two vectors have different dimensionalities, which shows that Hadamard multiplication is defined only for two vectors of equal dimensionality. You can fix the problem by removing one number from a or adding one number to b.

The outer product is quite different from the dot product: it produces a matrix instead of a scalar, and the two vectors in an outer product can have different dimensionalities, whereas the two vectors in a dot product must have the same dimensionality.

The outer product is indicated as $\mathbf{v}\mathbf{w}^T$ (remember that we assume vectors are in column orientation; therefore, the outer product involves multiplying a column by a row). Note the subtle but important difference between notation for the dot product ($\mathbf{v}^T\mathbf{w}$) and the outer product ($\mathbf{v}\mathbf{w}^T$). This might seem strange and confusing now, but I promise it will make perfect sense after learning about matrix multiplication in Chapter 5.

The outer product is similar to broadcasting, but they are not the same: *broadcasting* is a general coding operation that is used to expand vectors in arithmetic operations such as addition, multiplication, and division; the *outer product* is a specific mathematical procedure for multiplying two vectors.

NumPy can compute the outer product via the function `np.outer()` or the function `np.dot()` if the two input vectors are in, respectively, column and row orientation.

Cross and Triple Products

There are a few other ways to multiply vectors such as the cross product or triple product. Those methods are used in geometry and physics, but don't come up often enough in tech-related applications to spend any time on in this book. I mention them here only so you have passing familiarity with the names.

Orthogonal Vector Decomposition

To "decompose" a vector or matrix means to break up that matrix into multiple simpler pieces. Decompositions are used to reveal information that is "hidden" in a matrix, to make the matrix easier to work with, or for data compression. It is no understatement to write that much of linear algebra (in the abstract and in practice) involves matrix decompositions. Matrix decompositions are a big deal.

Let me introduce the concept of a decomposition using two simple examples with scalars:

- We can decompose the number 42.01 into two pieces: 42 and .01. Perhaps .01 is noise to be ignored, or perhaps the goal is to compress the data (the integer 42 requires less memory than the floating-point 42.01). Regardless of the motivation, the decomposition involves representing one mathematical object as the sum of simpler objects (42 = 42 + .01).

- We can decompose the number 42 into the product of prime numbers 2, 3, and 7. This decomposition is called *prime factorization* and has many applications in numerical processing and cryptography. This example involves products instead of sums, but the point is the same: decompose one mathematical object into smaller, simpler pieces.

In this section, we will begin exploring a simple yet important decomposition, which is to break up a vector into two separate vectors, one of which is orthogonal to a reference vector while the other is parallel to that reference vector. Orthogonal vector decomposition directly leads to the Gram-Schmidt procedure and QR decomposition, which is used frequently when solving inverse problems in statistics.

Let's begin with a picture so you can visualize the goal of the decomposition. Figure 2-6 illustrates the situation: we have two vectors **a** and **b** in standard position, and our goal is find the point on **a** that is as close as possible to the head of **b**. We could also express this as an optimization problem: project vector **b** onto vector **a** such that the projection distance is minimized. Of course, that point on **a** will be a scaled version of **a**; in other words, $\beta\mathbf{a}$. So now our goal is to find the scalar β. (The connection to orthogonal vector decomposition will soon be clear.)

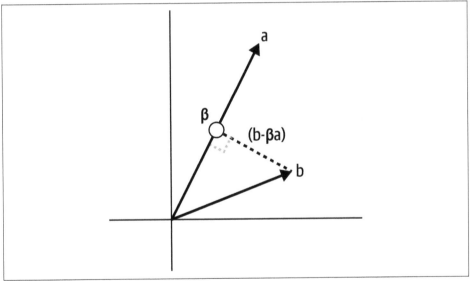

Figure 2-6. To project a point at the head of **b** *onto a vector* **a** *with minimum distance, we need a formula to compute* β *such that the length of the projection vector* $(\mathbf{b} - \beta\mathbf{a})$ *is minimized*

Importantly, we can use vector subtraction to define the line from **b** to $\beta\mathbf{a}$. We could give this line its own letter, e.g., vector **c**, but the subtraction is necessary for discovering the solution.

The key insight that leads to the solution to this problem is that the point on **a** that is closest to the head of **b** is found by drawing a line from **b** that meets **a** at a right angle. The intuition here is to imagine a triangle formed by the origin, the head of **b**, and β**a**; the length of the line from **b** to β**a** gets longer as the angle $\angle\beta$**a** gets smaller than 90° or larger than 90°.

Putting this together, we have deduced that (**b** – β**a**) is orthogonal to β**a**, which is the same thing as saying that those vectors are perpendicular. And that means that the dot product between them must be zero. Let's transform those words into an equation:

$$\mathbf{a}^T(\mathbf{b} - \beta\mathbf{a}) = 0$$

From here, we can apply some algebra to solve for β (note the application of the distributive property of dot products), which is shown in Equation 2-12:

Equation 2-12. Solving the orthogonal projection problem

$$\mathbf{a}^T\mathbf{b} - \beta\mathbf{a}^T\mathbf{a} = 0$$
$$\beta\mathbf{a}^T\mathbf{a} = \mathbf{a}^T\mathbf{b}$$
$$\beta = \frac{\mathbf{a}^T\mathbf{b}}{\mathbf{a}^T\mathbf{a}}$$

This is quite beautiful: we began with a simple geometric picture, explored the implications of the geometry, expressed those implications as a formula, and then applied a bit of algebra. And the upshot is that we discovered a formula for projecting a point onto a line with minimum distance. This is called *orthogonal projection*, and it is the basis for many applications in statistics and machine learning, including the famous least squares formula for solving linear models (you'll see orthogonal projections in Chapters 9, 10, and 11).

I can imagine that you're super curious to see what the Python code would look like to implement this formula. But you're going to have to write that code yourself in Exercise 2-8 at the end of this chapter. If you can't wait until the end of the chapter, feel free to solve that exercise now, and then continue learning about orthogonal decomposition.

You might be wondering how this is related to orthogonal vector decomposition, i.e., the title of this section. The minimum distance projection is the necessary grounding, and you're now ready to learn the decomposition.

As usual, we start with the setup and the goal. We begin with two vectors, which I'll call the "target vector" and the "reference vector." Our goal is to decompose the target vector into two other vectors such that (1) those two vectors sum to the target vector, and (2) one vector is orthogonal to the reference vector while the other is parallel to the reference vector. The situation is illustrated in Figure 2-7.

Before starting with the math, let's get our terms straight: I will call the target vector **t** and the reference vector **r**. Then, the two vectors formed from the target vector will be called the *perpendicular component*, indicated as $t_{\perp r}$, and the *parallel component*, indicated as $t_{\parallel r}$.

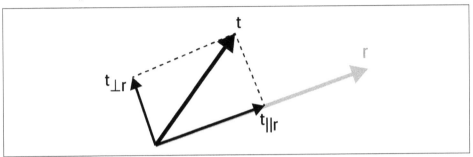

*Figure 2-7. Illustration of orthogonal vector decomposition: decompose vector **t** into the sum of two other vectors that are orthogonal and parallel to vector **r***

We begin by defining the parallel component. What is a vector that is parallel to **r**? Well, any scaled version of **r** is obviously parallel to **r**. So, we find $t_{\parallel r}$ simply by applying the orthogonal projection formula that we just discovered (Equation 2-13):

*Equation 2-13. Computing the parallel component of **t** with respect to **r***

$$t_{\parallel r} = r \frac{t^T r}{r^T r}$$

Note the subtle difference to Equation 2-12: there we only computed the scalar β; here we want to compute the scaled vector βr.

That's the parallel component. How do we find the perpendicular component? That one is easier, because we already know that the two vector components must sum to the original target vector. Thus:

$$t = t_{\perp r} + t_{\parallel r}$$
$$t_{\perp r} = t - t_{\parallel r}$$

In other words, we subtract off the parallel component from the original vector, and the residual is our perpendicular component.

But is that perpendicular component *really* orthogonal to the reference vector? Yes, it is! To prove it, you show that the dot product between the perpendicular component and the reference vector is zero:

$$\left(\mathbf{t}_{\perp \mathbf{r}}\right)^{\mathrm{T}} \mathbf{r} = 0$$

$$\left(\mathbf{t} - \mathbf{r} \frac{\mathbf{t}^{\mathrm{T}} \mathbf{r}}{\mathbf{r}^{\mathrm{T}} \mathbf{r}}\right)^{\mathrm{T}} \mathbf{r} = 0$$

Working through the algebra of this proof is straightforward but tedious, so I've omitted it. Instead, you'll work on building intuition using Python code in the exercises.

I hope you enjoyed learning about orthogonal vector decomposition. Note again the general principle: we break apart one mathematical object into a combination of other objects. The details of the decomposition depend on our constraints (in this case, orthogonal and parallel to a reference vector), which means that different constraints (that is, different goals of the analysis) can lead to different decompositions of the same vector.

Summary

The beauty of linear algebra is that even the most sophisticated and computationally intense operations on matrices are made up of simple operations, most of which can be understood with geometric intuition. Do not underestimate the importance of studying simple operations on vectors, because what you learned in this chapter will form the basis for the rest of the book—and the rest of your career as an *applied linear algebratician* (which is what you really are if you do anything with data science, machine learning, AI, deep learning, image processing, computational vision, statistics, blah blah blah).

Here are the most important take-home messages of this chapter:

- A vector is an ordered list of numbers that is placed in a column or in a row. The number of elements in a vector is called its dimensionality, and a vector can be represented as a line in a geometric space with the number of axes equal to the dimensionality.

- Several arithmetic operations (addition, subtraction, and Hadamard multiplication) on vectors work element-wise.

- The dot product is a single number that encodes the relationship between two vectors of the same dimensionality, and is computed as element-wise multiplication and sum.

- The dot product is zero for vectors that are orthogonal, which geometrically means that the vectors meet at a right angle.
- Orthogonal vector decomposition involves breaking up a vector into the sum of two other vectors that are orthogonal and parallel to a reference vector. The formula for this decomposition can be rederived from the geometry, but you should remember the phrase "mapping over magnitude" as the concept that that formula expresses.

Code Exercises

I hope you don't see these exercises as tedious work that you need to do. Instead, these exercises are opportunities to polish your math and coding skills, and to make sure that you really understand the material in this chapter.

I also want you to see these exercises as a springboard to continue exploring linear algebra using Python. Change the code to use different numbers, different dimensionalities, different orientations, etc. Write your own code to test other concepts mentioned in the chapter. Most importantly: have fun and embrace the learning experience.

As a reminder: the solutions to all the exercises can be viewed or downloaded from *https://github.com/mikexcohen/LA4DataScience*.

Exercise 2-1.

The online code repository is "missing" code to create Figure 2-2. (It's not really *missing*—I moved it into the solution to this exercise.) So, your goal here is to write your own code to produce Figure 2-2.

Exercise 2-2.

Write an algorithm that computes the norm of a vector by translating Equation 2-7 into code. Confirm, using random vectors with different dimensionalities and orientations, that you get the same result as `np.linalg.norm()`. This exercise is designed to give you more experience with indexing NumPy arrays and translating formulas into code; in practice, it's often easier to use `np.linalg.norm()`.

Exercise 2-3.

Create a Python function that will take a vector as input and output a unit vector in the same direction. What happens when you input the zeros vector?

Exercise 2-4.

You know how to create *unit* vectors; what if you want to create a vector of any arbitrary magnitude? Write a Python function that will take a vector and a desired magnitude as inputs and will return a vector in the same direction but with a magnitude corresponding to the second input.

Exercise 2-5.

Write a `for` loop to transpose a row vector into a column vector without using a built-in function or method such as `np.transpose()` or `v.T`. This exercise will help you create and index orientation-endowed vectors.

Exercise 2-6.

Here is an interesting fact: you can compute the squared norm of a vector as the dot product of that vector with itself. Look back to Equation 2-7 to convince yourself of this equivalence. Then confirm it using Python.

Exercise 2-7.

Write code to demonstrate that the dot product is *commutative*. Commutative means that $a \times b = b \times a$, which, for the vector dot product, means that $\mathbf{a}^{\mathsf{T}}\mathbf{b} = \mathbf{b}^{\mathsf{T}}\mathbf{a}$. After demonstrating this in code, use equation Equation 2-9 to understand why the dot product is commutative.

Exercise 2-8.

Write code to produce Figure 2-6. (Note that your solution doesn't need to look *exactly* like the figure, as long as the key elements are present.)

Exercise 2-9.

Implement orthogonal vector decomposition. Start with two random-number vectors $\mathbf{t}$ and $\mathbf{r}$, and reproduce Figure 2-8 (note that your plot will look somewhat different due to random numbers). Next, confirm that the two components sum to $\mathbf{t}$ and that $\mathbf{t}_{\perp \mathbf{r}}$ and $\mathbf{t}_{\parallel \mathbf{r}}$ are orthogonal.

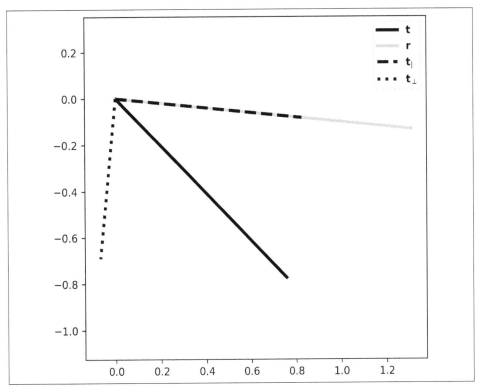

Figure 2-8. Exercise 9

Exercise 2-10.

An important skill in coding is finding bugs. Let's say there is a bug in your code such that the denominator in the projection scalar of Equation 2-13 is $\mathbf{t}^T\mathbf{t}$ instead of $\mathbf{r}^T\mathbf{r}$ (an easy mistake to make, speaking from personal experience while writing this chapter!). Implement this bug to check whether it really deviates from the accurate code. What can you do to check whether the result is correct or incorrect? (In coding, confirming code with known results is called *sanity-checking*.)

## CHAPTER 3

Vectors, Part 2

The previous chapter laid the groundwork for understanding vectors and basic operations acting on vectors. Now you will expand the horizons of your linear algebra knowledge by learning about a set of interrelated concepts including linear independence, subspaces, and bases. Each of these topics is crucially important for understanding operations on matrices.

Some of the topics here may seem abstract and disconnected from applications, but there is a very short path between them, e.g., vector subspaces and fitting statistical models to data. The applications in data science come later, so please keep focusing on the fundamentals so that the advanced topics are easier to understand.

Vector Sets

We can start the chapter with something easy: a collection of vectors is called a *set*. You can imagine putting a bunch of vectors into a bag to carry around. Vector sets are indicated using capital italics letters, like *S* or *V*. Mathematically, we can describe sets as the following:

$$V = \{\mathbf{v}_1, \ldots, \mathbf{v}_n\}$$

Imagine, for example, a dataset of the number of Covid-19 positive cases, hospitalizations, and deaths from one hundred countries; you could store the data from each country in a three-element vector, and create a vector set containing one hundred vectors.

Vector sets can contain a finite or an infinite number of vectors. Vector sets with an infinite number of vectors may sound like a uselessly silly abstraction, but vector subspaces are infinite vector sets and have major implications for fitting statistical models to data.

Vector sets can also be empty, and are indicated as $V = \{\}$. You'll encounter empty vector sets when you learn about matrix spaces.

Linear Weighted Combination

A *linear weighted combination* is a way of mixing information from multiple variables, with some variables contributing more than others. This fundamental operation is also sometimes called *linear mixture* or *weighted combination* (the *linear* part is assumed). Sometimes, the term *coefficient* is used instead of *weight*.

Linear weighted combination simply means scalar-vector multiplication and addition: take some set of vectors, multiply each vector by a scalar, and add them to produce a single vector (Equation 3-1).

> *Equation 3-1. Linear weighted combination*
>
> $\mathbf{w} = \lambda_1 \mathbf{v}_1 + \lambda_2 \mathbf{v}_2 + \ldots + \lambda_n \mathbf{v}_n$

It is assumed that all vectors $\mathbf{v}_i$ have the same dimensionality; otherwise, the addition is invalid. The λs can be any real number, including zero.

Technically, you could rewrite Equation 3-1 for subtracting vectors, but because subtraction can be handled by setting a λ_i to be negative, it's easier to discuss linear weighted combinations in terms of summation.

Equation 3-2 shows an example to help make it more concrete:

> *Equation 3-2. Linear weighted combination*
>
> $$\lambda_1 = 1, \ \lambda_2 = 2, \ \lambda_3 = -3, \quad \mathbf{v_1} = \begin{bmatrix} 4 \\ 5 \\ 1 \end{bmatrix}, \ \mathbf{v_2} = \begin{bmatrix} -4 \\ 0 \\ -4 \end{bmatrix}, \ \mathbf{v_3} = \begin{bmatrix} 1 \\ 3 \\ 2 \end{bmatrix}$$
>
> $$\mathbf{w} = \lambda_1 \mathbf{v_1} + \lambda_2 \mathbf{v_2} + \lambda_3 \mathbf{v_3} = \begin{bmatrix} -7 \\ -4 \\ -13 \end{bmatrix}$$

Linear weighted combinations are easy to implement, as the following code demonstrates. In Python, the data type is important; test what happens when the vectors are lists instead of NumPy arrays:[1]

---

[1] As shown in Chapters 2 and 16, list-integer multiplication repeats the list instead of scalar multiplying it.

```
l1 = 1
l2 = 2
l3 = -3
v1 = np.array([4,5,1])
v2 = np.array([-4,0,-4])
v3 = np.array([1,3,2])
l1*v1 + l2*v2 + l3*v3
```

Storing each vector and each coefficient as separate variables is tedious and does not scale up to larger problems. Therefore, in practice, linear weighted combinations are implemented via the compact and scalable matrix-vector multiplication method, which you'll learn about in Chapter 5; for now, the focus is on the concept and coding implementation.

Linear weighted combinations have several applications. Three of those include:

- The predicted data from a statistical model are created by taking the linear weighted combination of regressors (predictor variables) and coefficients (scalars) that are computed via the least squares algorithm, which you'll learn about in Chapters 11 and 12.

- In dimension-reduction procedures such as principal components analysis, each component (sometimes called factor or mode) is derived as a linear weighted combination of the data channels, with the weights (the coefficients) selected to maximize the variance of the component (along with some other constraints that you'll learn about in Chapter 15).

- Artificial neural networks (the architecture and algorithm that powers deep learning) involve two operations: linear weighted combination of the input data, followed by a nonlinear transformation. The weights are learned by minimizing a cost function, which is typically the difference between the model prediction and the real-world target variable.

The concept of a linear weighted combination is the mechanism of creating vector subspaces and matrix spaces, and is central to linear independence. Indeed, linear weighted combination and the dot product are two of the most important elementary building blocks from which many advanced linear algebra computations are built.

Linear Independence

A set of vectors is *linearly dependent* if at least one vector in the set can be expressed as a linear weighted combination of other vectors in that set. And thus, a set of vectors is *linearly independent* if no vector can be expressed as a linear weighted combination of other vectors in the set.

Following are two vector sets. Before reading the text, try to determine whether each set is dependent or independent. (The term *linear independence* is sometimes shortened to *independence* when the *linear* part is implied.)

$$V = \left\{ \begin{bmatrix} 1 \\ 3 \end{bmatrix}, \begin{bmatrix} 2 \\ 7 \end{bmatrix} \right\} \qquad S = \left\{ \begin{bmatrix} 1 \\ 3 \end{bmatrix}, \begin{bmatrix} 2 \\ 6 \end{bmatrix} \right\}$$

Vector set V is linearly indepedent: it is impossible to express one vector in the set as a linear multiple of the other vector in the set. That is to say, if we call the vectors in the set $\mathbf{v}_1$ and $\mathbf{v}_2$, then there is no possible scalar λ for which $\mathbf{v}_1 = \lambda \mathbf{v}_2$.

How about set S? This one is dependent, because we can use linear weighted combinations of some vectors in the set to obtain other vectors in the set. There is an infinite number of such combinations, two of which are $\mathbf{s}_1 = .5 * \mathbf{s}^*_2$ and $\mathbf{s}_2 = 2 * \mathbf{s}^*_1$.

Let's try another example. Again, the question is whether set T is linearly independent or linearly dependent:

$$T = \left\{ \begin{bmatrix} 8 \\ -4 \\ 14 \\ 6 \end{bmatrix}, \begin{bmatrix} 4 \\ 6 \\ 0 \\ 3 \end{bmatrix}, \begin{bmatrix} 14 \\ 2 \\ 4 \\ 7 \end{bmatrix}, \begin{bmatrix} 13 \\ 2 \\ 9 \\ 8 \end{bmatrix} \right\}$$

Wow, this one is a lot harder to figure out than the previous two examples. It turns out that this is a linearly dependent set (for example, the sum of the first three vectors equals twice the fourth vector). But I wouldn't expect you to be able to figure that out just from visual inspection.

So how do you determine linear independence in practice? The way to determine linear independence is to create a matrix from the vector set, compute the rank of the matrix, and compare the rank to the smaller of the number of rows or columns. That sentence may not make sense to you now, because you haven't yet learned about matrix rank. Therefore, focus your attention now on the concept that a set of vectors is linearly dependent if at least one vector in the set can be expressed as a linear weighted combination of the other vectors in the set, and a set of vectors is linearly independent if no vector can be expressed as a combination of other vectors.

Independent Sets

Independence is a property of a *set* of vectors. That is, a set of vectors can be linearly independent or linearly dependent; independence is not a property of an individual vector within a set.

The Math of Linear Independence

Now that you understand the concept, I want to make sure you also understand the formal mathematical definition of linear dependence, which is expressed in Equation 3-3.

Equation 3-3. Linear dependence[2]

$$\mathbf{0} = \lambda_1 \mathbf{v}_1 + \lambda_2 \mathbf{v}_2 + \ldots + \lambda_n \mathbf{v}_n, \quad \lambda \in \mathbb{R}$$

This equation says that linear dependence means that we can define some linear weighted combination of the vectors in the set to produce the zeros vector. If you can find some λs that make the equation true, then the set of vectors is linearly dependent. Conversely, if there is no possible way to linearly combine the vectors to produce the zeros vector, then the set is linearly independent.

That might initially be unintuitive. Why do we care about the zeros vector when the question is whether we can express at least one vector in the set as a weighted combination of other vectors in the set? Perhaps you'd prefer rewriting the definition of linear dependence as the following:

$$\lambda_1 \mathbf{v}_1 = \lambda_2 \mathbf{v}_2 + \ldots + \lambda_n \mathbf{v}_n, \quad \lambda \in \mathbb{R}$$

Why not start with that equation instead of putting the zeros vector on the left-hand side? Setting the equation to zero helps reinforce the principle that the *entire set* is dependent or independent; no individual vector has the privileged position of being the "dependent vector" (see "Independent Sets" on page 36). In other words, when it comes to independence, vector sets are purely egalitarian.

But wait a minute. Careful inspection of Equation 3-3 reveals a trivial solution: set all λ's to zero, and the equation reads $\mathbf{0} = \mathbf{0}$, regardless of the vectors in the set. But, as I wrote in Chapter 2, trivial solutions involving zeros are often ignored in linear algebra. So we add the constraint that at least one $\lambda \neq 0$.

This constraint can be incorporated into the equation by dividing through by one of the scalars; keep in mind that $\mathbf{v}_1$ and λ_1 can refer to any vector/scalar pair in the set:

$$\mathbf{0} = \mathbf{v}_1 + \ldots + \frac{\lambda_n}{\lambda_1} \mathbf{v}_n, \quad \lambda \in \mathbb{R}, \lambda_1 \neq 0$$

---

2 This equation is an application of linear weighted combination!

Independence and the Zeros Vector

Simply put, any vector set that includes the zeros vector is automatically a linearly dependent set. Here's why: any scalar multiple of the zeros vector is still the zeros vector, so the definition of linear dependence is always satisfied. You can see this in the following equation:

$$\lambda_0 \mathbf{0} = 0\mathbf{v}_1 + 0\mathbf{v}_2 + 0\mathbf{v}_n$$

As long as $\lambda_0 \neq 0$, we have a nontrivial solution, and the set fits with the definition of linear dependence.

What About Nonlinear Independence?

"But Mike," I imagine you protesting, "isn't life, the universe, and everything *non*linear?" I suppose it would be an interesting exercise to count the total number of linear versus nonlinear interactions in the universe and see which sum is larger. But linear algebra is all about, well, *linear* operations. If you can express one vector as a nonlinear (but not linear) combination of other vectors, then those vectors still form a linearly independent set. The reason for the linearity constraint is that we want to express transformations as matrix multiplication, which is a linear operation. That's not to throw shade on nonlinear operations—in my imaginary conversation, you have eloquently articulated that a purely linear universe would be rather dull and predictable. But we don't need to explain the entire universe using linear algebra; we need linear algebra only for the linear parts. (It's also worth mentioning that many nonlinear systems can be well approximated using linear functions.)

Subspace and Span

When I introduced linear weighted combinations, I gave examples with specific numerical values for the weights (e.g., $\lambda_1 = 1, \lambda_3 = -3$). A *subspace* is the same idea but using the infinity of possible ways to linearly combine the vectors in the set.

That is, for some (finite) set of vectors, the infinite number of ways to linearly combine them—using the same vectors but different numerical values for the weights —creates a *vector subspace*. And the mechanism of combining all possible linear weighted combinations is called the *span* of the vector set. Let's work through a few examples. We'll start with a simple example of a vector set containing one vector:

$$V = \left\{ \begin{bmatrix} 1 \\ 3 \end{bmatrix} \right\}$$

The span of this vector set is the infinity of vectors that can be created as linear combinations of the vectors in the set. For a set with one vector, that simply means all possible scaled versions of that vector. Figure 3-1 shows the vector and the subspace it spans. Consider that any vector in the gray dashed line can be formed as some scaled version of the vector.

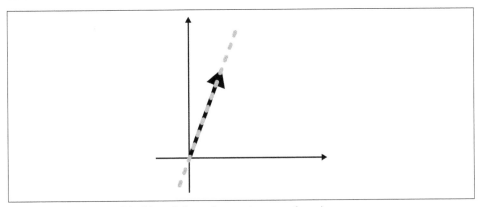

Figure 3-1. A vector (black) and the subspace it spans (gray)

Our next example is a set of two vectors in $\mathbb{R}^3$:

$$V = \left\{ \begin{bmatrix} 1 \\ 0 \\ 2 \end{bmatrix}, \begin{bmatrix} -1 \\ 1 \\ 2 \end{bmatrix} \right\}$$

The vectors are in $\mathbb{R}^3$, so they are graphically represented in a 3D axis. But the subspace that they span is a 2D plane in that 3D space (Figure 3-2). That plane passes through the origin, because scaling both vectors by zero gives the zeros vector.

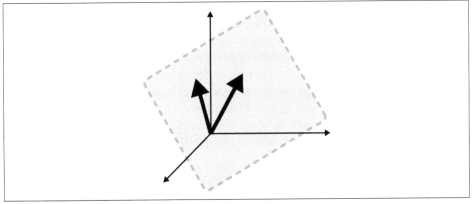

Figure 3-2. Two vectors (black) and the subspace they span (gray)

The first example had one vector and its span was a 1D subspace, and the second example had two vectors and their span was a 2D subspace. There seems to be a pattern emerging—but looks can be deceiving. Consider the next example:

$$V = \left\{ \begin{bmatrix} 1 \\ 1 \\ 1 \end{bmatrix}, \begin{bmatrix} 2 \\ 2 \\ 2 \end{bmatrix} \right\}$$

Two vectors in $\mathbb{R}^3$, but the subspace that they span is still only a 1D subspace—a line (Figure 3-3). Why is that? It's because one vector in the set is already in the span of the other vector. Thus, in terms of span, one of the two vectors is redundant.

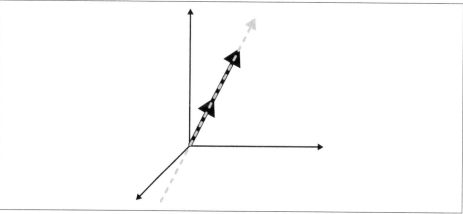

Figure 3-3. The 1D subspace (gray) spanned by two vectors (black)

So then, what is the relationship between the dimensionality of the spanned subspace and the number of vectors in the set? You might have guessed that it has something to do with linear independence.

The dimensionality of the subspace spanned by a set of vectors is the smallest number of vectors that forms a linearly independent set. If a vector set is linearly independent, then the dimensionality of the subspace spanned by the vectors in that set equals the number of vectors in that set. If the set is dependent, then the dimensionality of the subspace spanned by those vectors is necessarily less than the number of vectors in that set. Exactly how much smaller is another matter—to know the relationship between the number of vectors in a set and the dimensionality of their spanning subspace, you need to understand matrix rank, which you'll learn about in Chapter 6.

The formal definition of a vector subspace is a subset that is closed under addition and scalar multiplication, and includes the origin of the space. That means that any linear weighted combination of vectors in the subspace must also be in the same subspace, including setting all weights to zero to produce the zeros vector at the origin of the space.

Please don't lose sleep meditating on what it means to be "closed under addition and scalar multiplication"; just remember that a vector subspace is created from all possible linear combinations of a set of vectors.

What's the Difference Between Subspace and Span?

Many students are confused about the difference between *span* and *subspace*. That's understandable, because they are highly related concepts and often refer to the same thing. I will explain the difference between them, but don't stress about the subtleties—span and subspace so often refer to identical mathematical objects that using the terms interchangeably is usually correct.

I find that thinking of *span* as a verb and *subspace* as a noun helps understand their distinction: a set of vectors spans, and the result of their spanning is a subspace. Now consider that a *subspace* can be a smaller portion of a larger space, as you saw in Figure 3-3. Putting these together: span is the mechanism of creating a subspace. (On the other hand, when you use span as a noun, then span and subspace refer to the same infinite vector set.)

Basis

How far apart are Amsterdam and Tenerife? Approximately 2,000. What does "2,000" mean? That number makes sense only if we attach a basis unit. A basis is like a ruler for measuring a space.

In this example, the unit is *mile*. So our basis measurement for Dutch-Spanish distance is 1 mile. We could, of course, use different measurement units, like nanometers or light-years, but I think we can agree that mile is a convenient basis for distance at that scale. What about the length that your fingernail grows in one day—should we still use miles? Technically we can, but I think we can agree that *millimeter* is a more convenient basis unit. To be clear: the amount that your fingernail has grown in the past 24 hours is the same, regardless of whether you measure it in nanometers, miles, or light-years. But different units are more or less convenient for different problems.

Back to linear algebra: a *basis* is a set of rulers that you use to describe the information in the matrix (e.g., data). Like with the previous examples, you can describe the same data using different rulers, but some rulers are more convenient than others for solving certain problems.

The most common basis set is the Cartesian axis: the familiar XY plane that you've used since elementary school. We can write out the basis sets for the 2D and 3D Cartesian graphs as follows:

$$S_2 = \left\{ \begin{bmatrix} 1 \\ 0 \end{bmatrix}, \begin{bmatrix} 0 \\ 1 \end{bmatrix} \right\} \qquad S_3 = \left\{ \begin{bmatrix} 1 \\ 0 \\ 0 \end{bmatrix}, \begin{bmatrix} 0 \\ 1 \\ 0 \end{bmatrix}, \begin{bmatrix} 0 \\ 0 \\ 1 \end{bmatrix} \right\}$$

Notice that the Cartesian basis sets comprise vectors that are mutually orthogonal and unit length. Those are great properties to have, and that's why the Cartesian basis sets are so ubiquitous (indeed, they are called the *standard basis set*).

But those are not the only basis sets. The following set is a different basis set for $\mathbb{R}^2$.

$$T = \left\{ \begin{bmatrix} 3 \\ 1 \end{bmatrix}, \begin{bmatrix} -3 \\ 1 \end{bmatrix} \right\}$$

Basis set S_2 and T both span the same subspace (all of $\mathbb{R}^2$). Why would you prefer T over S? Imagine we want to describe data points p and q in Figure 3-4. We can describe those data points as their relationship to the origin—that is, their coordinates—using basis S or basis T.

In basis S, those two coordinates are $p = (3, 1)$ and $q = (-6, 2)$. In linear algebra, we say that the points are expressed as the linear combinations of the basis vectors. In this case, that combination is $3s_1 + 1s_2$ for point p, and $-6s_1 + 2s_2$ for point q.

Now let's describe those points in basis T. As coordinates, we have $p = (1, 0)$ and $q = (0, 2)$. And in terms of basis vectors, we have $1t_1 + 0t_2$ for point p and $0t_1 + 2t_2$ for point q (in other words, $p = t_1$ and $q = 2t_2$). Again, the data points p and q are the same regardless of the basis set, but T provided a compact and orthogonal description.

Bases are extremely important in data science and machine learning. In fact, many problems in applied linear algebra can be conceptualized as finding the best set of basis vectors to describe some subspace. You've probably heard of the following terms: dimension reduction, feature extraction, principal components analysis, independent components analysis, factor analysis, singular value decomposition, linear discriminant analysis, image approximation, data compression. Believe it or not, all of those analyses are essentially ways of identifying optimal basis vectors for a specific problem.

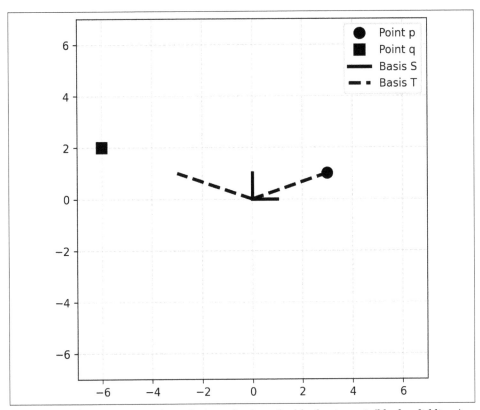

Figure 3-4. The same points (p and q) can be described by basis set S (black solid lines) or T (black dashed lines)

Consider Figure 3-5: this is a dataset of two variables (each dot represents a data point). The figure actually shows three distinct bases: the "standard basis set" corresponding to the x = 0 and y = 0 lines, and basis sets defined via a principal components analysis (PCA; left plot) and via an independent components analysis (ICA; right plot). Which of these basis sets provides the "best" way of describing the data? You might be tempted to say that the basis vectors computed from the ICA are the best. The truth is more complicated (as it tends to be): no basis set is intrinsically better or worse; different basis sets can be more or less helpful for specific problems based on the goals of the analysis, the features of the data, constraints imposed by the analyses, and so on.

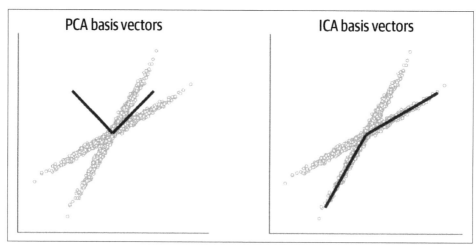

Figure 3-5. A 2D dataset using different basis vectors (black lines)

Definition of Basis

Once you understand the concept of a basis and basis set, the formal definition is straightforward. In fact, basis is simply the combination of span and independence: a set of vectors can be a basis for some subspace if it (1) spans that subspace and (2) is an independent set of vectors.

The basis needs to span a subspace for it to be used as a basis for that subspace, because you cannot describe something that you cannot measure.[3] Figure 3-6 shows an example of a point outside of a 1D subspace. A basis vector for that subspace cannot measure the point *r*. The black vector is still a valid basis vector for the subspace it spans, but it does not form a basis for any subspace beyond what it spans.

---

3 A general truism in science.

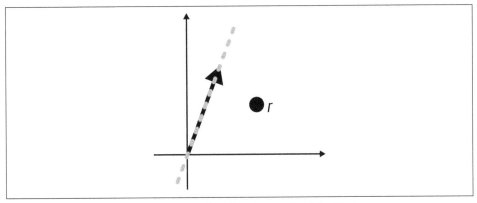

Figure 3-6. A basis set can measure only what is contained inside its span

So a basis needs to span the space that it is used for. That's clear. But why does a basis set require linear independence? The reason is that any given vector in the subspace must have a unique coordinate using that basis. Let's imagine describing point p from Figure 3-4 using the following vector set:

$$U = \left\{ \begin{bmatrix} 1 \\ 0 \end{bmatrix}, \begin{bmatrix} 2 \\ 0 \end{bmatrix}, \begin{bmatrix} 0 \\ 1 \end{bmatrix} \right\}$$

U is a perfectly valid vector set, but it is definitely *not* a basis set. Why not?[4]

What linear weighted combination describes point p in set U? Well, the coefficients for the linear weighted combination of the three vectors in U could be (3, 0, 1) or (0, 1.5, 1) or…a bajillion other possibilities. That's confusing, and so mathematicians decided that a vector must have *unique* coordinates within a basis set. Linear independence guarantees uniqueness.

To be clear, point p (or any other point) can be described using an infinite number of basis sets. So the measurement is not unique in terms of the plethora of possible basis sets. But *within* a basis set, a point is defined by exactly one linear weighted combination. It's the same thing with my distance analogy at the beginning of this section: we can measure the distance from Amsterdam to Tenerife using many different measurement units, but that distance has only one value per measurement unit. The distance is not simultaneously 3,200 miles and 2,000 miles, but it is simultaneously 3,200 *kilometers* and 2,000 *miles*. (Note for nerds: I'm approximating here, OK?)

---

4 Because it is a linearly dependent set.

Summary

Congratulations on finishing another chapter! (Well, almost finished: there are coding exercises to solve.) The point of this chapter was to bring your foundational knowledge about vectors to the next level. Below is a list of key points, but please remember that underlying all of these points is a very small number of elementary principles, primarily linear weighted combinations of vectors:

- A vector set is a collection of vectors. There can be a finite or an infinite number of vectors in a set.

- Linear weighted combination means to scalar multiply and add vectors in a set. Linear weighted combination is one of the single most important concepts in linear algebra.

- A set of vectors is linearly dependent if a vector in the set can be expressed as a linear weighted combination of other vectors in the set. And the set is linearly independent if there is no such linear weighted combination.

- A subspace is the infinite set of all possible linear weighted combinations of a set of vectors.

- A basis is a ruler for measuring a space. A vector set can be a basis for a subspace if it (1) spans that subspace and (2) is linearly independent. A major goal in data science is to discover the best basis set to describe datasets or to solve problems.

Code Exercises

Exercise 3-1.

Rewrite the code for linear weighted combination, but put the scalars in a list and the vectors as elements in a list (thus, you will have two lists, one of scalars and one of NumPy arrays). Then use a `for` loop to implement the linear weighted combination operation. Initialize the output vector using `np.zeros()`. Confirm that you get the same result as in the previous code.

Exercise 3-2.

Although the method of looping through lists in the previous exercise is not as efficient as matrix-vector multiplication, it is more scalable than without a `for` loop. You can explore this by adding additional scalars and vectors as elements in the lists. What happens if the new added vector is in $\mathbb{R}^4$ instead of $\mathbb{R}^3$? And what happens if you have more scalars than vectors?

Exercise 3-3.

In this exercise, you will draw random points in subspaces. This will help reinforce the idea that subspaces comprise *any* linear weighted combination of the spanning vectors. Define a vector set containing one vector [1, 3]. Then create 100 numbers drawn randomly from a uniform distribution between −4 and +4. Those are your random scalars. Multiply the random scalars by the basis vector to create 100 random points in the subspace. Plot those points.

Next, repeat the procedure but using two vectors in $\mathbb{R}^3$: [3, 5, 1] and [0, 2, 2]. Note that you need 100 × 2 random scalars for 100 points and two vectors. The resulting random dots will be on a plane. Figure 3-7 shows what the results will look like (it's not clear from the figure that the points lie on a plane, but you'll see this when you drag the plot around on your screen).

I recommend using the plotly library to draw the dots, so you can click-drag the 3D axis around. Here's a hint for getting it to work:

```
import plotly.graph_objects as go
fig = go.Figure( data=[go.Scatter3d(
                   x=points[:,0], y=points[:,1], z=points[:,2],
                   mode='markers' )])
fig.show()
```

Finally, repeat the $\mathbb{R}^3$ case but setting the second vector to be 1/2 times the first.

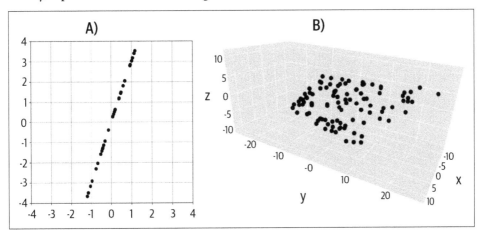

Figure 3-7. Exercise 3-3

## CHAPTER 4

Vector Applications

While working through the previous two chapters, you may have felt that some of the material was esoteric and abstract. Perhaps you felt that the challenge of learning linear algebra wouldn't pay off in understanding real-world applications in data science and machine learning.

I hope that this chapter dispels you of these doubts. In this chapter, you will learn how vectors and vector operations are used in data science analyses. And you will be able to extend this knowledge by working through the exercises.

Correlation and Cosine Similarity

Correlation is one of the most fundamental and important analysis methods in statistics and machine learning. A *correlation coefficient* is a single number that quantifies the linear relationship between two variables. Correlation coefficients range from −1 to +1, with −1 indicating a perfect negative relationship, +1 a perfect positive relationships, and 0 indicating no linear relationship. Figure 4-1 shows a few examples of pairs of variables and their correlation coefficients.

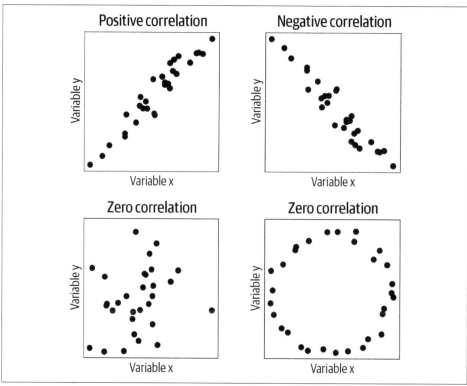

Figure 4-1. Examples of data exhibiting positive correlation, negative correlation, and zero correlation. The lower-right panel illustrates that correlation is a linear measure; nonlinear relationships between variables can exist even if their correlation is zero.

In Chapter 2, I mentioned that the dot product is involved in the correlation coefficient, and that the magnitude of the dot product is related to the magnitude of the numerical values in the data (remember the discussion about using grams versus pounds for measuring weight). Therefore, the correlation coefficient requires some normalizations to be in the expected range of −1 to +1. Those two normalizations are:

Mean center each variable
  Mean centering means to subtract the average value from each data value.

Divide the dot product by the product of the vector norms
  This divisive normalization cancels the measurement units and scales the maximum possible correlation magnitude to $|1|$.

Equation 4-1 shows the full formula for the Pearson correlation coefficient.

Equation 4-1. Formula for Pearson correlation coefficient

$$\rho = \frac{\sum_{i=1}^{n}(x_i - \bar{x})(y_i - \bar{y})}{\sqrt{\sum_{i=1}^{n}(x_i - \bar{x})^2}\sqrt{\sum_{i=1}^{n}(y_i - \bar{y})^2}}$$

It may not be obvious that the correlation is simply three dot products. Equation 4-2 shows this same formula rewritten using the linear algebra dot-product notation. In this equation, $\tilde{x}$ is the mean-centered version of $\mathbf{x}$ (that is, variable $\mathbf{x}$ with normalization #1 applied).

Equation 4-2. The Pearson correlation expressed in the parlance of linear algebra

$$\rho = \frac{\tilde{\mathbf{x}}^{\mathsf{T}}\tilde{\mathbf{y}}}{\|\tilde{\mathbf{x}}\| \; \|\tilde{\mathbf{y}}\|}$$

So there you go: the famous and widely used Pearson correlation coefficient is simply the dot product between two variables, normalized by the magnitudes of the variables. (By the way, you can also see from this formula that if the variables are unit normed such that $\|\mathbf{x}\| = \|\mathbf{y}\| = 1$, then their correlation equals their dot product. (Recall from Exercise 2-6 that $\|\mathbf{x}\| = \sqrt{\mathbf{x}^{\mathsf{T}}\mathbf{x}}$.)

Correlation is not the only way to assess similarity between two variables. Another method is called *cosine similarity*. The formula for cosine similarity is simply the geometric formula for the dot product (Equation 2-11), solved for the cosine term:

$$\cos\left(\theta_{x,y}\right) = \frac{\alpha}{\|\mathbf{x}\| \; \|\mathbf{y}\|}$$

where α is the dot product between $\mathbf{x}$ and $\mathbf{y}$.

It may seem like correlation and cosine similarity are exactly the same formula. However, remember that Equation 4-1 is the full formula, whereas Equation 4-2 is a simplification under the assumption that the variables have already been mean centered. Thus, cosine similarity does not involve the first normalization factor.

Correlation Versus Cosine Similarity

What does it mean for two variables to be "related"? Pearson correlation and cosine similarity can give different results for the same data because they start from different assumptions. In the eyes of Pearson, the variables [0, 1, 2, 3] and [100, 101, 102, 103] are perfectly correlated ($\rho = 1$) because changes in one variable are exactly mirrored in the other variable; it doesn't matter that one variable has larger numerical values. However, the cosine similarity between those variables is .808—they are not in the same numerical scale and are therefore not perfectly related. Neither measure is incorrect nor better than the other; it is simply the case that different statistical methods make different assumptions about data, and those assumptions have implications for the results—and for proper interpretation. You'll have the opportunity to explore this in Exercise 4-2.

From this section, you can understand why the Pearson correlation and cosine similarity reflects the *linear* relationship between two variables: they are based on the dot product, and the dot product is a linear operation.

There are four coding exercises associated with this section, which appear at the end of the chapter. You can choose whether you want to solve those exercises before reading the next section, or continue reading the rest of the chapter and then work through the exercises. (My personal recommendation is the former, but you are the master of your linear algebra destiny!)

Time Series Filtering and Feature Detection

The dot product is also used in time series filtering. Filtering is essentially a feature-detection method, whereby a template—called a *kernel* in the parlance of filtering—is matched against portions of a time series signal, and the result of filtering is another time series that indicates how much the characteristics of the signal match the characteristics of the kernel. Kernels are carefully constructed to optimize certain criteria, such as smooth fluctuations, sharp edges, particular waveform shapes, and so on.

The mechanism of filtering is to compute the dot product between the kernel and the time series signal. But filtering usually requires *local* feature detection, and the kernel is typically much shorter than the entire time series. Therefore, we compute the dot product between the kernel and a short snippet of the data of the same length as the kernel. This procedure produces one time point in the filtered signal (Figure 4-2), and then the kernel is moved one time step to the right to compute the dot product with a different (overlapping) signal segment. Formally, this procedure is called convolution and involves a few additional steps that I'm omitting to focus on the application of the dot product in signal processing.

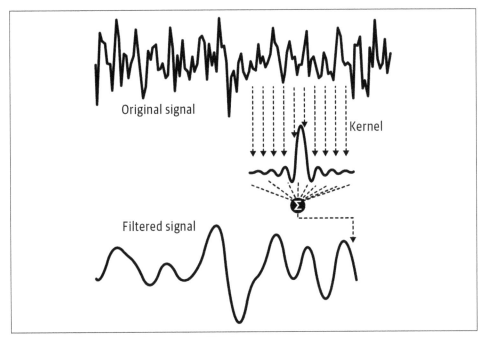

Figure 4-2. Illustration of time series filtering

Temporal filtering is a major topic in science and engineering. Indeed, without temporal filtering there would be no music, radio, telecommunications, satellites, etc. And yet, the mathematical heart that keeps your music pumping is the vector dot product.

In the exercises at the end of the chapter, you will discover how dot products are used to detect features (edges) and to smooth time series data.

k-Means Clustering

k-means clustering is an unsupervised method of classifying multivariate data into a relatively small number of groups, or categories, based on minimizing distance to the group center.

k-means clustering is an important analysis method in machine learning, and there are sophisticated variants of *k*-means clustering. Here we will implement a simple version of *k*-means, with the goal of seeing how concepts about vectors (in particular: vectors, vector norms, and broadcasting) are used in the *k*-means algorithm.

Here is a brief description of the algorithm that we will write:

1. Initialize k centroids as random points in the data space. Each centroid is a *class*, or category, and the next steps will assign each data observation to each class. (A *centroid* is a center generalized to any number of dimensions.)

2. Compute the Euclidean distance between each data observation and each centroid.[1]

3. Assign each data observation to the group with the closest centroid.

4. Update each centroid as the average of all data observations assigned to that centroid.

5. Repeat steps 2–4 until a convergence criteria is satisfied, or for N iterations.

If you are comfortable with Python coding and would like to implement this algorithm, then I encourage you to do that before continuing. Next, we will work through the math and code for each of these steps, with a particular focus on using vectors and broadcasting in NumPy. We will also test the algorithm using randomly generated 2D data to confirm that our code is correct.

Let's start with step 1: initialize k random cluster centroids. k is a parameter of k-means clustering; in real data, it is difficult to determine the optimal k, but here we will fix $k = 3$. There are several ways to initialize random cluster centroids; to keep things simple, I will randomly select k data samples to be centroids. The data are contained in variable data (this variable is 150×2, corresponding to 150 observations and 2 features) and visualized in the upper-left panel of Figure 4-3 (the online code shows how to generate these data):

```
k = 3
ridx = np.random.choice(range(len(data)),k,replace=False)
centroids = data[ridx,:] # data matrix is samples by features
```

Now for step 2: compute the distance between each data observation and each cluster centroid. Here is where we use linear algebra concepts you learned in the previous chapters. For one data observation and centroid, Euclidean distance is computed as

$$\delta_{i,j} = \sqrt{\left(d_i^x - c_j^x\right)^2 + \left(d_i^y - c_j^y\right)^2}$$

where $\delta_{i,j}$ indicates the distance from data observation i to centroid j, d^x_i is feature x of the ith data observation, and c^x_j is the x-axis coordinate of centroid j.

---

1 Reminder: Euclidean distance is the square root of the sum of squared distances from the data observation to the centroid.

You might think that this step needs to be implemented using a double `for` loop: one loop over k centroids and a second loop over N data observations (you might even think of a third `for` loop over data features). However, we can use vectors and broadcasting to make this operation compact and efficient. This is an example of how linear algebra often looks different in equations compared to in code:

```
dists = np.zeros((data.shape[0],k))
for ci in range(k):
    dists[:,ci] = np.sum((data-centroids[ci,:])**2,axis=1)
```

Let's think about the sizes of these variables: `data` is 150×2 (observations by features) and `centroids[ci,:]` is 1×2 (cluster `ci` by features). Formally, it is not possible to subtract these two vectors. However, Python will implement broadcasting by repeating the cluster centroids 150 times, thus subtracting the centroid from each data observation. The exponent operation `**` is applied element-wise, and the `axis=1` input tells Python to sum across the columns (separately per row). So, the output of `np.sum()` will be a 150×1 array that encodes the Euclidean distance of each point to centroid `ci`.

Take a moment to compare the code to the distance formula. Are they really the same? In fact, they are not: the square root in Euclidean distance is missing from the code. So is the code wrong? Think about this for a moment; I'll discuss the answer later.

Step 3 is to assign each data observation to the group with minimum distance. This step is quite compact in Python, and can be implemented using one function:

```
groupidx = np.argmin(dists,axis=1)
```

Note the difference between `np.min`, which returns the minimum *value*, versus `np.argmin`, which returns the *index* at which the minimum occurs.

We can now return to the inconsistency between the distance formula and its code implementation. For our k-means algorithm, we use distance to assign each data point to its closest centroid. Distance and squared distance are monotonically related, so both metrics give the same answer. Adding the square root operation increases code complexity and computation time with no impact on the results, so it can simply be omitted.

Step 4 is to recompute the centroids as the mean of all data points within the class. Here we can loop over the k clusters, and use Python indexing to find all data points assigned to each cluster:

```
for ki in range(k):
    centroids[ki,:] = [ np.mean(data[groupidx==ki,0]),
                        np.mean(data[groupidx==ki,1])  ]
```

Finally, Step 5 is to put the previous steps into a loop that iterates until a good solution is obtained. In production-level *k*-means algorithms, the iterations continue until a stopping criteria is reached, e.g., that the cluster centroids are no longer moving around. For simplicity, here we will iterate three times (an arbitrary number selected to make the plot visually balanced).

The four panels in Figure 4-3 show the initial random cluster centroids (iteration 0), and their updated locations after each of three iterations.

If you study clustering algorithms, you will learn sophisticated methods for centroid initialization and stopping criteria, as well as quantitative methods to select an appropriate *k* parameter. Nonetheless, all *k*-means methods are essentially extensions of the above algorithm, and linear algebra is at the heart of their implementations.

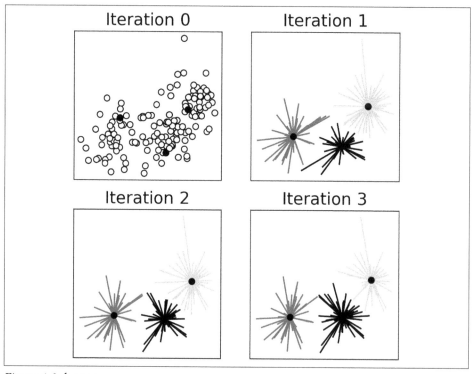

Figure 4-3. k-means

Code Exercises

Correlation Exercises

Exercise 4-1.

Write a Python function that takes two vectors as input and provides two numbers as output: the Pearson correlation coefficient and the cosine similarity value. Write code that follows the formulas presented in this chapter; don't simply call `np.corrcoef` and `spatial.distance.cosine`. Check that the two output values are identical when the variables are already mean centered and different when the variables are not mean centered.

Exercise 4-2.

Let's continue exploring the difference between correlation and cosine similarity. Create a variable containing the integers 0 through 3, and a second variable equaling the first variable plus some offset. You will then create a simulation in which you systematically vary that offset between −50 and +50 (that is, the first iteration of the simulation will have the second variable equal to [−50, −49, −48, −47]). In a `for` loop, compute the correlation and cosine similarity between the two variables and store these results. Then make a line plot showing how the correlation and cosine similarity are affected by the mean offset. You should be able to reproduce Figure 4-4.

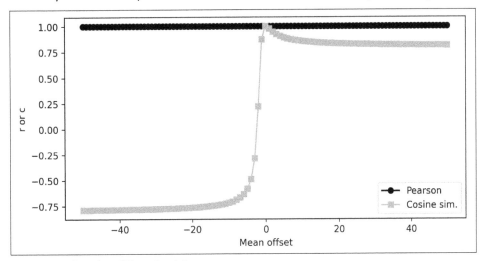

Figure 4-4. Results of Exercise 4-2

Exercise 4-3.

There are several Python functions to compute the Pearson correlation coefficient. One of them is called `pearsonr` and is located in the `stats` module of the SciPy library. Open the source code for this file (hint: `??functionname`) and make sure you understand how the Python implementation maps onto the formulas introduced in this chapter.

Exercise 4-4.

Why do you ever need to code your own functions when they already exist in Python? Part of the reason is that writing your own functions has huge educational value, because you see that (in this case) the correlation is a simple computation and not some incredibly sophisticated black-box algorithm that only a computer-science PhD could understand. But another reason is that built-in functions are sometimes slower because of myriad input checks, dealing with additional input options, converting data types, etc. This increases usability but at the expense of computation time.

Your goal in this exercise is to determine whether your own bare-bones correlation function is faster than NumPy's `corrcoef` function. Modify the function from Exercise 4-2 to compute only the correlation coefficient. Then, in a `for` loop over 1,000 iterations, generate two variables of 500 random numbers and compute the correlation between them. Time the `for` loop. Then repeat but using `np.corrcoef`. In my tests, the custom function was about 33% faster than `np.corrcoef`. In these toy examples, the differences are measured in milliseconds, but if you are running billions of correlations with large datasets, those milliseconds really add up! (Note that writing your own functions without input checks has the risk of input errors that would be caught by `np.corrcoef`.) (Also note that the speed advantage breaks down for larger vectors. Try it!)

Filtering and Feature Detection Exercises

Exercise 4-5.

Let's build an edge detector. The kernel for an edge detector is very simple: $[-1 +1]$. The dot product of that kernel with a snippet of a time series signal with constant value (e.g., $[10\ 10]$) is 0. But that dot product is large when the signal has a steep change (e.g., $[1\ 10]$ would produce a dot product of 9). The signal we'll work with is a plateau function. Graphs A and B in Figure 4-5 show the kernel and the signal. The first step in this exercise is to write code that creates these two time series.

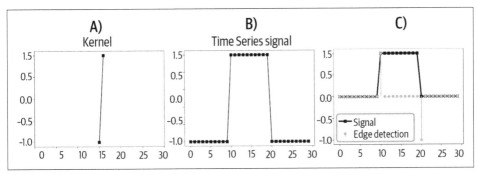

Figure 4-5. Results of Exercise 4-5

Next, write a for loop over the time points in the signal. At each time point, compute the dot product between the kernel and a segment of the time series data that has the same length as the kernel. You should produce a plot that looks like graph C in Figure 4-5. (Focus more on the result than on the aesthetics.) Notice that our edge detector returned 0 when the signal was flat, +1 when the signal jumped up, and −1 when the signal jumped down.

Feel free to continue exploring this code. For example, does anything change if you pad the kernel with zeros ([0 −1 1 0])? What about if you flip the kernel to be [1 −1]? How about if the kernel is asymmetric ([−1 2])?

Exercise 4-6.

Now we will repeat the same procedure but with a different signal and kernel. The goal will be to smooth a rugged time series. The time series will be 100 random numbers generated from a Gaussian distribution (also called a normal distribution). The kernel will be a bell-shaped function that approximates a Gaussian function, defined as the numbers [0, .1, .3, .8, 1, .8, .3, .1, 0] but scaled so that the sum over the kernel is 1. Your kernel should match graph A in Figure 4-6, although your signal won't look exactly like graph B due to random numbers.

Copy and adapt the code from the previous exercise to compute the sliding time series of dot products—the signal filtered by the Gaussian kernel. Warning: be mindful of the indexing in the for loop. Graph C in Figure 4-6 shows an example result. You can see that the filtered signal is a smoothed version of the original signal. This is also called low-pass filtering.

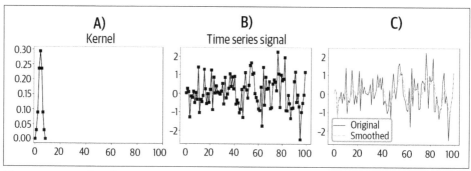

Figure 4-6. Results of Exercise 4-6

Exercise 4-7.

Replace the 1 in the center of the kernel with −1 and mean center the kernel. Then rerun the filtering and plotting code. What is the result? It actually accentuates the sharp features! In fact, this kernel is now a high-pass filter, meaning it dampens the smooth (low-frequency) features and highlights the rapidly changing (high-frequency) features.

k-Means Exercises

Exercise 4-8.

One way to determine an optimal k is to repeat the clustering multiple times (each time using randomly initialized cluster centroids) and assess whether the final clustering is the same or different. Without generating new data, rerun the k-means code several times using $k = 3$ to see whether the resulting clusters are similar (this is a qualitative assessment based on visual inspection). Do the final cluster assignments generally seem similar even though the centroids are randomly selected?

Exercise 4-9.

Repeat the multiple clusterings using $k = 2$ and $k = 4$. What do you think of these results?

## CHAPTER 5

Matrices, Part 1

A matrix is a vector taken to the next level. Matrices are highly versatile mathematical objects. They can store sets of equations, geometric transformations, the positions of particles over time, financial records, and myriad other things. In data science, matrices are sometimes called data tables, in which rows correspond to observations (e.g., customers) and columns correspond to features (e.g., purchases).

This and the following two chapters will take your knowledge about linear algebra to the next level. Get a cup of coffee and put on your thinking cap. Your brain will be bigger by the end of the chapter.

Creating and Visualizing Matrices in NumPy

Depending on the context, matrices can be conceptualized as a set of column vectors stacked next to each other (e.g., a data table of observations-by-features), as a set of row vectors layered on top of each other (e.g., multisensor data in which each row is a time series from a different channel), or as an ordered collection of individual matrix elements (e.g., an image in which each matrix element encodes pixel intensity value).

Visualizing, Indexing, and Slicing Matrices

Small matrices can simply be printed out in full, like the following examples:

$$\begin{bmatrix} 1 & 2 \\ \pi & 4 \\ 6 & 7 \end{bmatrix}, \quad \begin{bmatrix} -6 & 1/3 \\ e^{4.3} & -1.4 \\ 6/5 & 0 \end{bmatrix}$$

But that's not scalable, and matrices that you work with in practice can be large, perhaps containing billions of elements. Therefore, larger matrices can be visualized as images. The numerical value of each element of the matrix maps onto a color in the image. In most cases, the maps are pseudo-colored because the mapping of numerical value onto color is arbitrary. Figure 5-1 shows some examples of matrices visualized as images using the Python library `matplotlib`.

Figure 5-1. Three matrices, visualized as images

Matrices are indicated using bold-faced capital letters, like matrix **A** or **M**. The size of a matrix is indicated using (row, column) convention. For example, the following matrix is 3 × 5 because it has three rows and five columns:

$$\begin{bmatrix} 1 & 3 & 5 & 7 & 9 \\ 0 & 2 & 4 & 6 & 8 \\ 1 & 4 & 7 & 8 & 9 \end{bmatrix}$$

You can refer to specific elements of a matrix by indexing the row and column position: the element in the 3rd row and 4th column of matrix **A** is indicated as $a_{3,4}$ (and in the previous example matrix, $a_{3,4} = 8$). *Important reminder:* math uses 1-based indexing whereas Python uses 0-based indexing. Thus, element $a_{3,4}$ is indexed in Python as `A[2,3]`.

Extracting a subset of rows or columns of a matrix is done through slicing. If you are new to Python, you can consult Chapter 16 for an introduction to slicing lists and NumPy arrays. To extract a section out of a matrix, you specify the start and end rows and columns, and that the slicing goes in steps of 1. The online code walks you through the procedure, and the following code shows an example of extracting a submatrix from rows 2–4 and columns 1–5 of a larger matrix:

```
A = np.arange(60).reshape(6,10)
sub = A[1:4:1,0:5:1]
```

And here are the full and submatrices:

```
Original matrix:
[[ 0  1  2  3  4  5  6  7  8  9]
 [10 11 12 13 14 15 16 17 18 19]
 [20 21 22 23 24 25 26 27 28 29]
```

```
 [30 31 32 33 34 35 36 37 38 39]
 [40 41 42 43 44 45 46 47 48 49]
 [50 51 52 53 54 55 56 57 58 59]]

Submatrix:
[[10 11 12 13 14]
 [20 21 22 23 24]
 [30 31 32 33 34]]
```

Special Matrices

There is an infinite number of matrices, because there is an infinite number of ways of organizing numbers into a matrix. But matrices can be described using a relatively small number of characteristics, which creates "families" or categories of matrices. These categories are important to know, because they appear in certain operations or have certain useful properties.

Some categories of matrices are used so frequently that they have dedicated NumPy functions to create them. Following is a list of some common special matrices and Python code to create them;[1] you can see what they look like in Figure 5-2:

Random numbers matrix
> This is a matrix that contains numbers drawn at random from some distribution, typically Gaussian (a.k.a. normal). Random-numbers matrices are great for exploring linear algebra in code, because they are quickly and easily created with any size and rank (matrix rank is a concept you'll learn about in Chapter 16).
>
> There are several ways to create random matrices in NumPy, depending on which distribution you want to draw numbers from. In this book, we'll mostly use Gaussian-distributed numbers:
>
> ```
> Mrows = 4 # shape 0
> Ncols = 6 # shape 1
> A = np.random.randn(Mrows,Ncols)
> ```

Square versus nonsquare
> A square matrix has the same number of rows as columns; in other words, the matrix is in $\mathbb{R}^{N \times N}$. A nonsquare matrix, also sometimes called a rectangular matrix, has a different number of rows and columns. You can create square and rectangular matrices from random numbers by adjusting the shape parameters in the previous code.
>
> Rectangular matrices are called *tall* if they have more rows than columns and *wide* if they have more columns than rows.

---

1 There are more special matrices that you will learn about later in the book, but this list is enough to get started.

Diagonal

The *diagonal* of a matrix is the elements starting at the top-left and going down to the bottom-right. A *diagonal matrix* has zeros on all the off-diagonal elements; the diagonal elements may also contain zeros, but they are the only elements that may contain nonzero values.

The NumPy function `np.diag()` has two behaviors depending on the inputs: input a matrix and `np.diag` will return the diagonal elements as a vector; input a vector and `np.diag` will return a matrix with those vector elements on the diagonal. (Note: extracting the diagonal elements of a matrix is *not* called "diagonalizing a matrix"; that is a separate operation introduced in Chapter 13.)

Triangular

A triangular matrix contains all zeros either above or below the main diagonal. The matrix is called *upper triangular* if the nonzero elements are above the diagonal and *lower triangular* if the nonzero elements are below the diagonal.

NumPy has dedicated functions to extract the upper (`np.triu()`) or lower (`np.tril()`) triangle of a matrix.

Identity

The identity matrix is one of the most important special matrices. It is the equivalent of the number 1, in that any matrix or vector times the identity matrix is that same matrix or vector. The identity matrix is a square diagonal matrix with all diagonal elements having a value of 1. It is indicated using the letter **I**. You might see a subscript to indicate its size (e.g., I_5 is the 5×5 identity matrix); if not, then you can infer the size from context (e.g., to make the equation consistent).

You can create an identity matrix in Python using `np.eye()`.

Zeros

The zeros matrix is comparable to the zeros vector: it is the matrix of all zeros. Like the zeros vector, it is indicated using a bold-faced zero: **0**. It can be a bit confusing to have the same symbol indicate both a vector and a matrix, but this kind of overloading is common in math and science notation.

The zeros matrix is created using the `np.zeros()` function.

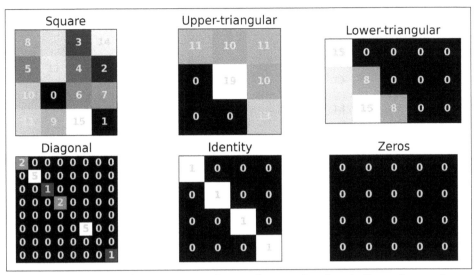

Figure 5-2. Some special matrices. Numbers and grayscale values indicate the matrix value at each element.

Matrix Math: Addition, Scalar Multiplication, Hadamard Multiplication

Mathematical operations on matrices fall into two categories: intuitive and unintuitive. In general, the intuitive operations can be expressed as element-wise procedures whereas the unintuitive operations take longer to explain and a bit of practice to understand. Let's start with the intuitive operations.

Addition and Subtraction

You add two matrices by adding their corresponding elements. Here's an example:

$$\begin{bmatrix} 2 & 3 & 4 \\ 1 & 2 & 4 \end{bmatrix} + \begin{bmatrix} 0 & 3 & 1 \\ -1 & -4 & 2 \end{bmatrix} = \begin{bmatrix} (2+0) & (3+3) & (4+1) \\ (1-1) & (2-4) & (4+2) \end{bmatrix} = \begin{bmatrix} 2 & 6 & 5 \\ 0 & -2 & 6 \end{bmatrix}$$

As you might guess from the example, matrix addition is defined only between two matrices of the same size.

"Shifting" a Matrix

As with vectors, it is not formally possible to add a scalar to a matrix, as in $\lambda + \mathbf{A}$. Python allows such an operation (e.g., `3+np.eye(2)`), which involves broadcast-adding the scalar to each element of the matrix. That is a convenient computation, but it is not formally a linear algebra operation.

But there is a linear-algebra way to add a scalar to a square matrix, and that is called *shifting* a matrix. It works by adding a constant value to the diagonal, which is implemented by adding a scalar multiplied identity matrix:

$$\mathbf{A} + \lambda \mathbf{I}$$

Here's a numerical example:

$$\begin{bmatrix} 4 & 5 & 1 \\ 0 & 1 & 11 \\ 4 & 9 & 7 \end{bmatrix} + 6 \begin{bmatrix} 1 & 0 & 0 \\ 0 & 1 & 0 \\ 0 & 0 & 1 \end{bmatrix} = \begin{bmatrix} 10 & 5 & 1 \\ 0 & 7 & 11 \\ 4 & 9 & 13 \end{bmatrix}$$

Shifting in Python is straightforward:

```
A = np.array([ [4,5,1],[0,1,11],[4,9,7] ])
s = 6
A + s # NOT shifting!
A + s*np.eye(len(A)) # shifting
```

Notice that only the diagonal elements change; the rest of the matrix is unadulterated by shifting. In practice, one shifts a relatively small amount to preserve as much information as possible in the matrix while benefiting from the effects of shifting, including increasing the numerical stability of the matrix (you'll learn later in the book why that happens).

Exactly how much to shift is a matter of ongoing research in multiple areas of machine learning, statistics, deep learning, control engineering, etc. For example, is shifting by $\lambda = 6$ a little or a lot? How about $\lambda = .001$? Obviously, these numbers are "big" or "small" relative to the numerical values in the matrix. Therefore, in practice, λ is usually set to be some fraction of a matrix-defined quantity such as the norm or the average of the eigenvalues. You'll get to explore this in later chapters.

"Shifting" a matrix has two primary (extremely important!) applications: it is the mechanism of finding the eigenvalues of a matrix, and it is the mechanism of regularizing matrices when fitting models to data.

Scalar and Hadamard Multiplications

These two types of multiplication work the same for matrices as they do for vectors, which is to say, element-wise.

Scalar-matrix multiplication means to multiply each element in the matrix by the same scalar. Here is an example using a matrix comprising letters instead of numbers:

$$\gamma \begin{bmatrix} a & b \\ c & d \end{bmatrix} = \begin{bmatrix} \gamma a & \gamma b \\ \gamma c & \gamma d \end{bmatrix}$$

Likewise, Hadamard multiplication involves multiplying two matrices element-wise (hence the alternative terminology *element-wise multiplication*). Here is an example:

$$\begin{bmatrix} 2 & 3 \\ 4 & 5 \end{bmatrix} \odot \begin{bmatrix} a & b \\ c & d \end{bmatrix} = \begin{bmatrix} 2a & 3b \\ 4c & 5d \end{bmatrix}$$

In NumPy, Hadamard multiplication can be implemented using the `np.multiply()` function. But it's often easier to implement using an asterisk between the two matrices: `A*B`. This can cause some confusion, because standard matrix multiplication (next section) is indicated using an @ symbol. That's a subtle but important difference! (This will be particularly confusing for readers coming to Python from MATLAB, where * indicates matrix multiplication.)

```
A = np.random.randn(3,4)
B = np.random.randn(3,4)

A*B # Hadamard multiplication
np.multiply(A,B) # also Hadamard
A@B # NOT Hadamard!
```

Hadamard multiplication does have some applications in linear algebra, for example, when computing the matrix inverse. However, it is most often used in applications as a convenient way to store many individual multiplications. That's similar to how vector Hadamard multiplication is often used, as discussed in Chapter 2.

Standard Matrix Multiplication

Now we get to the unintuitive way to multiply matrices. To be clear, standard matrix multiplication is not particularly difficult; it's just different from what you might expect. Rather than operating element-wise, standard matrix multiplication operates row/column-wise. In fact, standard matrix multiplication reduces to a systematic collection of dot products between rows of one matrix and columns of the other matrix. (This form of multiplication is formally simply called *matrix multiplication*;

I've added the term *standard* to help disambiguate from Hadamard and scalar multiplications.)

But before I get into the details of how to multiply two matrices, I will first explain how to determine whether two matrices can be multiplied. As you'll learn, two matrices can be multiplied only if their sizes are concordant.

Rules for Matrix Multiplication Validity

You know that matrix sizes are written out as $M \times N$—rows by columns. Two matrices multiplying each other can have different sizes, so let's refer to the size of the second matrix as $N \times K$. When we write out the two multiplicand matrices with their sizes underneath, we can refer to the "inner" dimensions N and the "outer" dimensions M and K.

Here's the important point: *matrix multiplication is valid only when the "inner" dimensions match, and the size of the product matrix is defined by the "outer" dimensions.* See Figure 5-3.

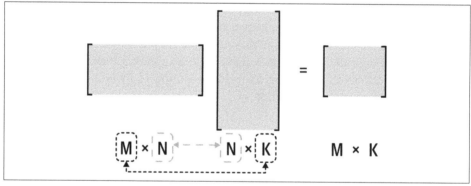

Figure 5-3. Matrix multiplication validity, visualized. Memorize this picture.

More formally, matrix multiplication is valid when the number of columns in the left matrix equals the number of rows in the right matrix, and the size of the product matrix is defined by the number of rows in the left matrix and the number of columns in the right matrix. I find the "inner/outer" rubric easier to remember.

You can already see that matrix multiplication does not obey the commutative law: **AB** may be valid while **BA** is invalid. Even if both multiplications are valid (for example, if both matrices are square), they may produce different results. That is, if **C = AB** and **D = BA**, then in general **C ≠ D** (they are equal in some special cases, but we cannot generally assume equality).

Note the notation: Hadamard multiplication is indicated using a dotted-circle ($\mathbf{A} \odot \mathbf{B}$) whereas matrix multiplication is indicated as two matrices side-by-side without any symbol between them ($\mathbf{AB}$).

Now it's time to learn about the mechanics and interpretation of matrix multiplication.

Matrix Multiplication

The reason why matrix multiplication is valid only if the number of columns in the left matrix matches the number of rows in the right matrix is that the (i,j)th element in the product matrix is the dot product between the ith row of the left matrix and the jth column in the right matrix.

Equation 5-1 shows an example of matrix multiplication, using the same two matrices that we used for Hadamard multiplication. Make sure you understand how each element in the product matrix is computed as dot products of corresponding rows and columns of the left-hand side matrices.

Equation 5-1. Example of matrix multiplication. Parentheses added to facilitate visual grouping.

$$\begin{bmatrix} 2 & 3 \\ 4 & 5 \end{bmatrix}\begin{bmatrix} a & b \\ c & d \end{bmatrix} = \begin{bmatrix} (2a + 3c) & (2b + 3d) \\ (4a + 5c) & (4b + 5d) \end{bmatrix}$$

If you are struggling to remember how matrix multiplication works, Figure 5-4 shows a mnemonic trick for drawing out the multiplication with your fingers.

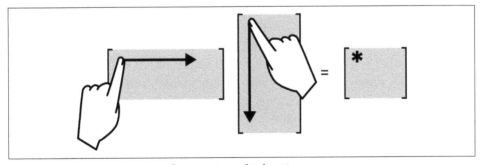

Figure 5-4. Finger movements for matrix multiplication

How do you interpret matrix multiplication? Remember that the dot product is a number that encodes the relationship between two vectors. So, the result of matrix multiplication is a matrix that stores all the pairwise linear relationships between rows of the left matrix and columns of the right matrix. That is a beautiful thing, and is the basis for computing covariance and correlation matrices, the general linear model (used in statistics analyses including ANOVAs and regressions), singular-value decomposition, and countless other applications.

Matrix-Vector Multiplication

In a purely mechanical sense, matrix-vector multiplication is nothing special and does not deserve its own subsection: mutliplying a matrix and a vector is simply matrix multiplication where one "matrix" is a vector.

But matrix-vector multiplication does have many applications in data science, machine learning, and computer graphics, so it's worth spending some time on. Let's start with the basics:

- A matrix can be right-multiplied by a column vector but not a row vector, and it can be left-multiplied by a row vector but not a column vector. In other words, $\mathbf{Av}$ and $\mathbf{v}^T\mathbf{A}$ are valid, but $\mathbf{Av}^T$ and $\mathbf{vA}$ are invalid.

  That is clear from inspecting matrix sizes: an $M \times N$ matrix can be premultiplied by a $1 \times M$ matrix (a.k.a. a row vector) or postmultiplied by an $N \times 1$ matrix (a.k.a. a column vector).

- The result of matrix-vector multiplication is always a vector, and the orientation of that vector depends on the orientation of the multiplicand vector: premultiplying a matrix by a row vector produces another row vector, while postmultiplying a matrix by a column vector produces another column vector. Again, this is obvious when you think about matrix sizes, but it's worth pointing out.

Matrix-vector multiplication has several applications. In statistics, the model-predicted data values are obtained by multiplying the design matrix by the regression coefficients, which is written out as $\mathbf{X}\boldsymbol{\beta}$. In principal components analysis, a vector of "feature-importance" weights is identified that maximizes variance in dataset $\mathbf{Y}$, and is written out as $\left(\mathbf{Y}^T\mathbf{Y}\right)\mathbf{v}$ (that feature-importance vector $\mathbf{v}$ is called an eigenvector). In multivariate signal processing, a reduced-dimensional component is obtained by applying a spatial filter to multichannel time series data $\mathbf{S}$, and is written out as $\mathbf{w}^T\mathbf{S}$. In geometry and computer graphics, a set of image coordinates can be transformed using a mathematical transformation matrix, and is written out as $\mathbf{Tp}$, where $\mathbf{T}$ is the transformation matrix and $\mathbf{p}$ is the set of geometric coordinates.

There are so many more examples of how matrix-vector multiplication is used in applied linear algebra, and you will see several of these examples later in the book. Matrix-vector multiplication is also the basis for matrix spaces, which is an important topic that you'll learn about later in the next chapter.

For now, I want to focus on two specific interpretations of matrix-vector multiplication: as a means to implement linear weighted combinations of vectors, and as the mechanism of implementing geometric transformations.

Linear weighted combinations

In the previous chapter, we calculated linear weighted combinations by having separate scalars and vectors, and then multiplying them individually. But you are now smarter than when you started the previous chapter, and so you are now ready to learn a better, more compact, and more scalable method for computing linear weighted combinations: put the individual vectors into a matrix, and put the weights into corresponding elements of a vector. Then multiply. Here's a numerical example:

$$4\begin{bmatrix}3\\0\\6\end{bmatrix} + 3\begin{bmatrix}1\\2\\5\end{bmatrix} \quad\Rightarrow\quad \begin{bmatrix}3 & 1\\0 & 2\\6 & 5\end{bmatrix}\begin{bmatrix}4\\3\end{bmatrix}$$

Please take a moment to work through the multiplication, and make sure you understand how the linear weighted combination of the two vectors can be implemented as a matrix-vector multiplication. The key insight is that each element in the vector scalar multiplies the corresponding column in the matrix, and then the weighted column vectors are summed to obtain the product.

This example involved linear weighted combinations of column vectors; what would you change to compute linear weighted combinations of row vectors?[2]

Geometric transforms

When we think of a vector as a geometric line, then matrix-vector multiplication becomes a way of rotating and scaling that vector (remember that scalar-vector multiplication can scale but not rotate).

Let's start with a 2D case for easy visualization. Here are our matrix and vectors:

```
M  = np.array([ [2,3],[2,1] ])
x  = np.array([ [1,1.5] ]).T
Mx = M@x
```

Notice that I created x as a row vector and then transposed it into a column vector; that reduced the number of square brackets to type.

Graph A in Figure 5-5 visualizes these two vectors. You can see that the matrix M both rotated and stretched the original vector. Let's try a different vector with the same matrix. Actually, just for fun, let's use the same vector elements but with swapped positions (thus, vector v = [1.5,1]).

Now a strange thing happens in graph B (Figure 5-5): the matrix-vector product is no longer rotated into a different direction. The matrix still scaled the vector, but its

---

2 Put the coefficients into a row vector and premultiply.

direction was preserved. In other words, the *matrix*-vector multiplication acted as if it were *scalar*-vector multiplication. That is not a random event: in fact, vector v is an eigenvector of matrix M, and the amount by which M stretched v is its eigenvalue. That's such an incredibly important phenomenon that it deserves its own chapter (Chapter 13), but I just couldn't resist introducing you to the concept now.

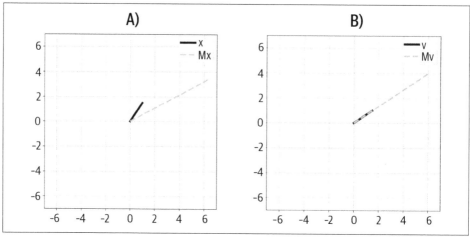

Figure 5-5. Examples of matrix-vector multiplication

Segues to advanced topics aside, the main point of these demonstrations is that one of the functions of matrix-vector multiplication is for a matrix to house a transformation that, when applied to a vector, can rotate and stretch that vector.

Matrix Operations: Transpose

You learned about the transpose operation on vectors in Chapter 2. The principle is the same with matrices: swap the rows and columns. And just like with vectors, the transpose is indicated with a superscripted T (thus, $\mathbf{C}^T$ is the transpose of $\mathbf{C}$). And double-transposing a matrix returns the original matrix ($\mathbf{C}^{TT} = \mathbf{C}$).

The formal mathematical definition of the transpose operation is printed in Equation 5-2 (essentially repeated from the previous chapter), but I think it's just as easy to remember that *transposing swaps rows and columns*.

Equation 5-2. Definition of the transpose operation

$$a_{i,j}^T = a_{j,i}$$

Here's an example:

$$\begin{bmatrix} 3 & 0 & 4 \\ 9 & 8 & 3 \end{bmatrix}^{T} = \begin{bmatrix} 3 & 9 \\ 0 & 8 \\ 4 & 3 \end{bmatrix}$$

There are a few ways to transpose matrices in Python, using a function and a method acting on NumPy arrays:

```
A = np.array([ [3,4,5],[1,2,3] ])
A_T1 = A.T # as method
A_T2 = np.transpose(A) # as function
```

The matrix in this example uses a 2D NumPy array; what do you think will happen if you apply the transpose method to a vector encoded as a 1D array? Try it and find out![3]

Dot and Outer Product Notation

Now that you know about the transpose operation and about the rules for matrix multiplication validity, we can return to the notation of the vector dot product. For two-column vectors of $M \times 1$, transposing the first vector and not the second gives two "matrices" of sizes $1 \times M$ and $M \times 1$. The "inner" dimensions match and the "outer" dimensions tell us that the product will be 1×1, a.k.a. a scalar. That is the reason why the dot product is indicated as $\mathbf{a}^{T}\mathbf{b}$.

Same reasoning for the outer product: multiplying a column vector by a row vector has sizes $M \times 1$ and $1 \times N$. The "inner" dimensions match, and the size of the result will be $M \times N$.

Matrix Operations: LIVE EVIL (Order of Operations)

LIVE EVIL is a palindrome (a palindrome is a word or phrase that is spelled the same forwards and backwards) and a cute mnemonic for remembering how transposing affects the order of multiplied matrices. Basically, the rule is that the transpose of multiplied matrices is the same as the individual matrices transposed and multiplied, but reversed in order. In Equation 5-3, **L**, **I**, **V**, and **E** are all matrices, and you can assume that their sizes match to make multiplication valid.

Equation 5-3. Example of the LIVE EVIL rule

$$(\mathbf{LIVE})^{T} = \mathbf{E}^{T}\mathbf{V}^{T}\mathbf{I}^{T}\mathbf{L}^{T}$$

---

[3] Nothing. NumPy will return the same 1D array without altering it, and without giving a warning or error.

Needless to say, this rule applies for multiplying any number of matrices, not just four, and not just with these "randomly selected" letters.

This does seem like a strange rule, but it's the only way to make transposing multiplied matrices work. You'll have the opportunity to test this yourself in Exercise 5-7 at the end of this chapter. If you like, you may skip to that exercise before moving on.

Symmetric Matrices

Symmetric matrices have lots of special properties that make them great to work with. They also tend to be numerically stable and thus convenient for computer algorithms. You'll learn about the special properties of symmetric matrices as you work through this book; here I will focus on what symmetric matrices are and how to create them from nonsymmetric matrices.

What does it mean for a matrix to be symmetric? It means that the corresponding rows and columns are equal. And that means that when you swap the rows and columns, nothing happens to the matrix. And that in turn means that a *symmetric matrix equals its transpose*. In math terms, a matrix $\mathbf{A}$ is symmetric if $\mathbf{A}^T = \mathbf{A}$.

Check out the symmetric matrix in Equation 5-4.

Equation 5-4. A symmetric matrix; note that each row equals its corresponding column

$$\begin{bmatrix} a & e & f & g \\ e & b & h & i \\ f & h & c & j \\ g & i & j & d \end{bmatrix}$$

Can a nonsquare matrix be symmetric? Nope! The reason is that if a matrix is of size $M \times N$, then its transpose is of size $N \times M$. Those two matrices could not be equal unless $M = N$, which means the matrix is square.

Creating Symmetric Matrices from Nonsymmetric Matrices

This may be surprising at first, but multiplying *any* matrix—even a nonsquare and nonsymmetric matrix—by its transpose will produce a square symmetric matrix. In other words, $\mathbf{A}^T\mathbf{A}$ is square symmetric, as is $\mathbf{A}\mathbf{A}^T$. (If you lack the time, patience, or keyboard skills to format the superscripted ^{T}, you can write AtA and AAt or A′A and AA′.)

Let's prove this claim rigorously before seeing an example. On the one hand, we don't actually need to prove separately that $\mathbf{A}^T\mathbf{A}$ is square *and* symmetric, because the latter implies the former. But proving squareness is straightforward and a good exercise in linear algebra proofs (which tend to be shorter and easier than, e.g., calculus proofs).

The proof is obtained simply by considering the matrix sizes: if $\mathbf{A}$ is $M \times N$, then $\mathbf{A}^T\mathbf{A}$ is $(N \times M)(M \times N)$, which means the product matrix is of size $N \times N$. You can work through the same logic for $\mathbf{A}\mathbf{A}^T$.

Now to prove symmetry. Recall that the definition of a symmetric matrix is one that equals its transpose. So let's transpose $\mathbf{A}^T\mathbf{A}$, do some algebra, and see what happens. Make sure you can follow each step here; the proof relies on the LIVE EVIL rule:

$$\left(\mathbf{A}^T\mathbf{A}\right)^T = \mathbf{A}^T\mathbf{A}^{TT} = \mathbf{A}^T\mathbf{A}$$

Taking the first and final terms, we get $\left(\mathbf{A}^T\mathbf{A}\right)^T = \left(\mathbf{A}^T\mathbf{A}\right)$. The matrix equals its transpose, hence it is symmetric.

Now repeat the proof on your own using $\mathbf{A}\mathbf{A}^T$. Spoiler alert! You'll come to the same conclusion. But writing out the proof will help you internalize the concept.

So $\mathbf{A}\mathbf{A}^T$ and $\mathbf{A}^T\mathbf{A}$ are both square symmetric. But they are not the same matrix! In fact, if $\mathbf{A}$ is nonsquare, then the two matrix products are not even the same size.

$\mathbf{A}^T\mathbf{A}$ is called the *multiplicative method* for creating symmetric matrices. There is also an *additive method*, which is valid when the matrix is square but nonsymmetric. This method has some interesting properties but doesn't have a lot of application value, so I won't focus on it. Exercise 5-9 walks you through the algorithm; if you're up for a challenge, you can try to discover that algorithm on your own before looking at the exercise.

Summary

This chapter is the first of a three-chapter series on matrices. Here you learned the groundwork from which all matrix operations are based. In summary:

- Matrices are spreadsheets of numbers. In different applications, it is useful to conceptualize them as a set of column vectors, a set of row vectors, or an arrangement of individual values. Regardless, visualizing matrices as images is often insightful, or at least pleasant to look at.

- There are several categories of special matrices. Being familiar with the properties of the types of matrices will help you understand matrix equations and advanced applications.

- Some arithmetic operations work element-wise, like addition, scalar multiplication, and Hadamard multiplication.

- "Shifting" a matrix means adding a constant to the diagonal elements (without changing the off-diagonal elements). Shifting has several applications in machine learning, primarily for finding eigenvalues and regularizing statistical models.

- Matrix multiplication involves dot products between rows of the left matrix and columns of the right matrix. The product matrix is an organized collection of mappings between row-column pairs. Memorize the rule for matrix multiplication validity: $(M \times N)(N \times K) = (M \times K)$.

- LIVE EVIL:[4] The transpose of multiplied matrices equals the individual matrices transposed and multiplied with their order reversed.

- Symmetric matrices are mirrored across the diagonal, which means that each row equals its corresponding columns, and are defined as $\mathbf{A} = \mathbf{A}^\mathrm{T}$. Symmetric matrices have many interesting and useful properties that make them great to work with in applications.

- You can create a symmetric matrix from any matrix by multiplying that matrix by its transpose. The resulting matrix $\mathbf{A}^\mathrm{T}\mathbf{A}$ is central to statistical models and the singular value decomposition.

Code Exercises

Exercise 5-1.

This exercise will help you gain familiarity with indexing matrix elements. Create a 3×4 matrix using `np.arange(12).reshape(3,4)`. Then write Python code to extract the element in the second row, fourth column. Use softcoding so that you can select different row/column indices. Print out a message like the following:

```
The matrix element at index (2,4) is 7.
```

Exercise 5-2.

This and the following exercise focus on slicing matrices to obtain submatrices. Start by creating matrix $\mathbf{C}$ in Figure 5-6, and use Python slicing to extract the submatrix comprising the first five rows and five columns. Let's call this matrix $\mathbf{C}_1$. Try to reproduce Figure 5-6, but if you are struggling with the Python visualization coding, then just focus on extracting the submatrix correctly.

---

4 LIVE EVIL is a cute mnemonic, not a recommendation for how to behave in society!

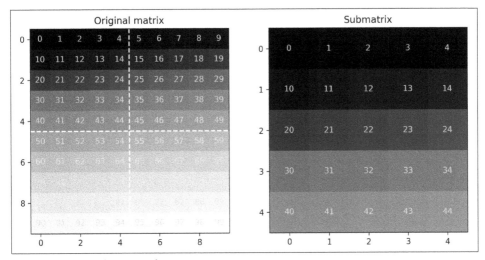

Figure 5-6. Visualization of Exercise 5-2

Exercise 5-3.

Expand this code to extract the other four 5 × 5 blocks. Then create a new matrix with those blocks reorganized according to Figure 5-7.

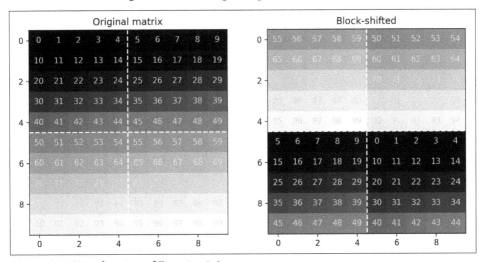

Figure 5-7. Visualization of Exercise 5-3

Exercise 5-4.

Implement matrix addition element-wise using two for loops over rows and columns. What happens when you try to add two matrices with mismatching sizes? This exercise will help you think about breaking down a matrix into rows, columns, and individual elements.

Exercise 5-5.

Matrix addition and scalar multiplication obey the mathematical laws of commutivity and distributivity. That means that the following equations give the same results (assume that the matrices **A** and **B** are the same size and that σ is some scalar):

$$\sigma(\mathbf{A} + \mathbf{B}) = \sigma\mathbf{A} + \sigma\mathbf{B} = \mathbf{A}\sigma + \mathbf{B}\sigma$$

Rather than proving this mathematically, you are going to demonstrate it through coding. In Python, create two random-numbers matrices of size 3×4 and a random scalar. Then implement the three expressions in the previous equation. You'll need to figure out a way to confirm that the three results are equal. Keep in mind that tiny computer precision errors in the range of 10^{-15} should be ignored.

Exercise 5-6.

Code matrix multiplication using `for` loops. Confirm your results against using the numpy @ operator. This exercise will help you solidify your understanding of matrix multiplication, but in practice, it's always better to use @ instead of writing out a double `for` loop.

Exercise 5-7.

Confirm the LIVE EVIL rule using the following five steps: (1) Create four matrices of random numbers, setting the sizes to be $\mathbf{L} \in \mathbb{R}^{2 \times 6}$, $\mathbf{I} \in \mathbb{R}^{6 \times 3}$, $\mathbf{V} \in \mathbb{R}^{3 \times 5}$, and $\mathbf{E} \in \mathbb{R}^{5 \times 2}$. (2) Multiply the four matrices and transpose the product. (3) Transpose each matrix individually and multiply them *without reversing their order*. (4) Transpose each matrix individually and multiply them reversing their order *according to the LIVE EVIL rule*. Check whether the result of step 2 matches the results of step 3 and step 4. (5) Repeat the previous steps but using all square matrices.

Exercise 5-8.

In this exercise, you will write a Python function that checks whether a matrix is symmetric. It should take a matrix as input, and should output a boolean `True` if the matrix is symmetric or `False` if the matrix is nonsymmetric. Keep in mind that small computer rounding/precision errors can make "equal" matrices appear unequal.

Therefore, you will need to test for equality with some reasonable tolerance. Test the function on symmetric and nonsymmetric matrices.

Exercise 5-9.

I mentioned that there is an additive method for creating a symmetric matrix from a nonsymmetric square matrix. The method is quite simple: average the matrix with its transpose. Implement this algorithm in Python and confirm that the result really is symmetric. (Hint: you can use the function you wrote in the previous exercise!)

Exercise 5-10.

Repeat the second part of Exercise 3-3 (the two vectors in $\mathbb{R}^3$), but use matrix-vector multiplication instead of vector-scalar multiplication. That is, compute $\mathbf{As}$ instead of $\sigma_1 \mathbf{v}_1 + \sigma_2 \mathbf{v}_2$.

Exercise 5-11.

Diagonal matrices have many interesting properties that make them useful to work with. In this exercise, you will learn about two of those properties:

- Premultiplying by a diagonal matrix scales the rows of the right matrix by the corresponding diagonal elements.
- Postmultiplying by a diagonal matrix scales the columns of the left matrix by the corresponding diagonal elements.

This fact is used in several applications, including computing correlation matrices (Chapter 7) and diagonalizing a matrix (Chapters 13 and 14).

Let's explore an implication of this property. Start by creating three 4×4 matrices: a matrix of all ones (hint: `np.ones()`); a diagonal matrix where the diagonal elements are 1, 4, 9, and 16; and a diagonal matrix equal to the square root of the previous diagonal matrix.

Next, print out the pre- and postmultiplied ones matrix by the first diagonal matrix. You'll get the following results:

```
# Pre-multiply by diagonal:
[[ 1.  1.  1.   1.]
 [ 4.  4.  4.   4.]
 [ 9.  9.  9.   9.]
 [16. 16. 16.  16.]]

# Post-multiply by diagonal:
[[ 1.  4.  9.  16.]
 [ 1.  4.  9.  16.]
```

```
[ 1.  4.  9. 16.]
[ 1.  4.  9. 16.]]
```

Finally, premultiply *and* postmultiply the ones matrix by the square root of the diagonal matrix. You'll get the following:

```
# Pre- and post-multiply by sqrt-diagonal:
[[ 1.  2.  3.  4.]
 [ 2.  4.  6.  8.]
 [ 3.  6.  9. 12.]
 [ 4.  8. 12. 16.]]
```

Notice that the rows *and* the columns are scaled such that the (i,j)th element in the matrix is multiplied by the product of the ith and jth diagonal elements. (In fact, we've created a multiplication table!)

Exercise 5-12.

Another fun fact: matrix multiplication is the same thing as Hadamard multiplication for two diagonal matrices. Figure out why this is using paper and pencil with two 3×3 diagonal matrices, and then illustrate it in Python code.

## CHAPTER 6

Matrices, Part 2

Matrix multiplication is one of the most wonderful gifts that mathematicians have bestowed upon us. But to move from elementary to advanced linear algebra—and then to understand and develop data science algorithms—you need to do more than just multiply matrices.

We begin this chapter with discussions of matrix norms and matrix spaces. Matrix norms are essentially an extension of vector norms, and matrix spaces are essentially an extension of vector subspaces (which in turn are nothing more than linear weighted combinations). So you already have the necessary background knowledge for this chapter.

Concepts like linear independence, rank, and determinant will allow you to transition from understanding elementary concepts like transpose and multiplication to understanding advanced topics like inverse, eigenvalues, and singular values. And those advanced topics unlock the power of linear algebra for applications in data science. Therefore, this chapter is a waypoint in your transformation from linear algebra newbie to linear algebra knowbie.[1]

Matrices seem like such simple things—just a spreadsheet of numbers. But you've already seen in the previous chapters that there is more to matrices than meets the eye. So, take a deep and calming breath and dive right in.

---

[1] The internet claims this is a real word; let's see if I can get it past the O'Reilly editors.

Matrix Norms

You learned about vector norms in Chapter 2: the norm of a vector is its Euclidean geometric length, which is computed as the square root of the sum of the squared vector elements.

Matrix norms are a little more complicated. For one thing, there is no "*the* matrix norm"; there are multiple distinct norms that can be computed from a matrix. Matrix norms are somewhat similar to vector norms in that each norm provides one number that characterizes a matrix, and that the norm is indicated using double-vertical lines, as in the norm of matrix $\mathbf{A}$ is indicated as $\| \mathbf{A} \|$.

But different matrix norms have different meanings. The myriad of matrix norms can be broadly divided into two families: element-wise (also sometimes called entry-wise) and induced. Element-wise norms are computed based on the individual elements of the matrix, and thus these norms can be interpreted to reflect the magnitudes of the elements in the matrix.

Induced norms can be interpreted in the following way: one of the functions of a matrix is to encode a transformation of a vector; the induced norm of a matrix is a measure of how much that transformation scales (stretches or shrinks) that vector. This interpretation will make more sense in Chapter 7 when you learn about using matrices for geometric transformations, and in Chapter 14 when you learn about the singular value decomposition.

In this chapter, I will introduce you to element-wise norms. I'll start with the Euclidean norm, which is actually a direct extension of the vector norm to matrices. The Euclidean norm is also called the *Frobenius norm*, and is computed as the square root of the sum of all matrix elements squared (Equation 6-1).

Equation 6-1. The Frobenius norm

$$\| \mathbf{A} \|_F = \sqrt{\sum_{i=1}^{M} \sum_{j=1}^{N} a_{ij}^2}$$

The indices i and j correspond to the M rows and N columns. Also note the subscripted $_F$ indicating the Frobenius norm.

The Frobenius norm is also called the $\ell 2$ norm (the ℓ is a fancy-looking letter L). And the $\ell 2$ norm gets its name from the general formula for element-wise p-norms (notice that you get the Frobenius norm when $p = 2$):

$$\| \mathbf{A} \|_p = \left(\sum_{i=1}^{M} \sum_{j=1}^{N} |a_{ij}|^p \right)^{1/p}$$

Matrix norms have several applications in machine learning and statistical analysis. One of the important applications is in regularization, which aims to improve model fitting and increase generalization of models to unseen data (you'll see examples of this later in the book). The basic idea of regularization is to add a matrix norm as a cost function to a minimization algorithm. That norm will help prevent model parameters from becoming too large ($\ell 2$ reguarlization, also called *ridge regression*) or encouraging sparse solutions ($\ell 1$ regularization, also called *lasso regression*). In fact, modern deep learning architectures rely on matrix norms to achieve such impressive performance at solving computer vision problems.

Another application of the Frobenius norm is computing a measure of "matrix distance." The distance between a matrix and itself is 0, and the distance between two distinct matrices increases as the numerical values in those matrices become increasingly dissimilar. Frobenius matrix distance is computed simply by replacing matrix **A** with matrix **C** = **A** − **B** in Equation 6-1.

This distance can be used as an optimization criterion in machine learning algorithms, for example, to reduce the data storage size of an image while minimizing the Frobenius distance between the reduced and original matrices. Exercise 6-2 will guide you through a simple minimization example.

Matrix Trace and Frobenius Norm

The *trace* of a matrix is the sum of its diagonal elements, indicated as $tr(\mathbf{A})$, and exists only for square matrices. Both of the following matrices have the same trace (14):

$$\begin{bmatrix} 4 & 5 & 6 \\ 0 & 1 & 4 \\ 9 & 9 & 9 \end{bmatrix}, \begin{bmatrix} 0 & 0 & 0 \\ 0 & 8 & 0 \\ 1 & 2 & 6 \end{bmatrix}$$

Trace has some interesting properties. For example, the trace of a matrix equals the sum of its eigenvalues and therefore is a measure of the "volume" of its eigenspace. Many properties of the trace are less relevant for data science applications, but here is one interesting exception:

$$\| A \|_F = \sqrt{\sum_{i=1}^{M} \sum_{j=1}^{N} a_{ij}^2} = \sqrt{tr(\mathbf{A}^T \mathbf{A})}$$

In other words, the Frobenius norm can be calculated as the square root of the trace of the matrix times its transpose. The reason why this works is that each diagonal element of the matrix $\mathbf{A}^T\mathbf{A}$ is defined by the dot product of each row with itself.

Exercise 6-3 will help you explore the trace method of computing the Frobenius norm.

Matrix Spaces (Column, Row, Nulls)

The concept of *matrix spaces* is central to many topics in abstract and applied linear algebra. Fortunately, matrix spaces are conceptually straightforward, and are essentially just linear weighted combinations of different features of a matrix.

Column Space

Remember that a linear weighted combination of vectors involves scalar multiplying and summing a set of vectors. Two modifications to this concept will extend linear weighted combination to the column space of a matrix. First, we conceptualize a matrix as a set of column vectors. Second, we consider the infinity of real-valued scalars instead of working with a specific set of scalars. An infinite number of scalars gives an infinite number of ways to combine a set of vectors. That resulting infinite set of vectors is called the *column space of a matrix.*

Let's make this concrete with some numerical examples. We'll start simple—a matrix that has only one column (which is actually the same thing as a column vector). Its column space—all possible linear weighted combinations of that column—can be expressed like this:

$$C\left(\begin{bmatrix}1\\3\end{bmatrix}\right) = \lambda\begin{bmatrix}1\\3\end{bmatrix}, \quad \lambda \in \mathbb{R}$$

The $C(\mathbf{A})$ indicates the column space of matrix $\mathbf{A}$, and the $\in$ symbol means "is a member of" or "is contained in." In this context, it means that λ can be any possible real-valued number.

What does this mathematical expression mean? It means that the column space is the set of all possible scaled versions of the column vector [1 3]. Let's consider a few specific cases. Is the vector [1 3] in the column space? Yes, because you can express that vector as the matrix times $\lambda = 1$. How about [−2 −6]? Also yes, because you can express that vector as the matrix times $\lambda = -2$. How about [1 4]? The answer here is no: vector [1 4] is *not* in the column space of the matrix, because there is simply no scalar that can multiply the matrix to produce that vector.

What does the column space look like? For a matrix with one column, the column space is a line that passes through the origin, in the direction of the column vector, and stretches out to infinity in both directions. (Technically, the line doesn't stretch out to literal infinity, because infinity is not a real number. But that line is arbitrarily long—much longer than our limited human minds can possibly fathom—so for all intents and purposes, we can speak of that line as being infinitely long.) Figure 6-1 shows a picture of the column space for this matrix.

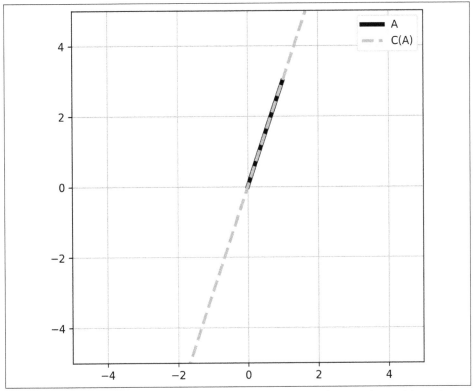

Figure 6-1. Visualization of the column space of a matrix with one column. This column space is a 1D subspace.

Now let's consider a matrix with more columns. We'll keep the column dimensionality to two so we can visualize it on a 2D graph. Here's our matrix and its column space:

$$C\left(\begin{bmatrix} 1 & 1 \\ 3 & 2 \end{bmatrix}\right) = \lambda_1 \begin{bmatrix} 1 \\ 3 \end{bmatrix} + \lambda_2 \begin{bmatrix} 1 \\ 2 \end{bmatrix}, \quad \lambda \in \mathbb{R}$$

We have two columns, so we allow for two distinct λs (they're both real-valued numbers but can be different from each other). Now the question is, what is the set of all vectors that can be reached by some linear combination of these two column vectors?

The answer is: all vectors in $\mathbb{R}^2$. For example, the vector $[-4\ 3]$ can be obtained by scaling the two columns by, respectively, 11 and −15. How did I come up with those scalar values? I used the least squares projection method, which you'll learn about in Chapter 11. For now, you can focus on the concept that these two columns can be appropriately weighted to reach any point in $\mathbb{R}^2$.

Graph A in Figure 6-2 shows the two matrix columns. I didn't draw the column space of the matrix because it is the entirety of the axis.

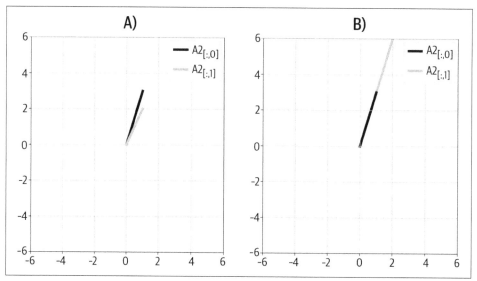

Figure 6-2. More examples of column spaces

One more example in $\mathbb{R}^2$. Here's our new matrix for consideration:

$$C\left(\begin{bmatrix} 1 & 2 \\ 3 & 6 \end{bmatrix}\right) = \lambda_1 \begin{bmatrix} 1 \\ 3 \end{bmatrix} + \lambda_2 \begin{bmatrix} 2 \\ 6 \end{bmatrix}, \quad \lambda \in \mathbb{R}$$

What is the dimensionality of its column space? Is it possible to reach any point in $\mathbb{R}^2$ by some linear weighted combination of the two columns?

The answer to the second question is no. And if you don't believe me, try to find a linear weighted combination of the two columns that produces vector [3 5]. It is simply impossible. In fact, the two columns are collinear (graph B in Figure 6-2), because one is already a scaled version of the other. That means that the column space of this 2×2 matrix is still just a line—a 1D subspace.

The take-home message here is that having N columns in a matrix does not guarantee that the column space will be N-D. The dimensionality of the column space equals the number of columns only if the columns form a linearly independent set. (Remember from Chapter 3 that linear independence means a set of vectors in which no vector can be expressed as a linear weighted combination of other vectors in that set.)

One final example of column spaces, to see what happens when we move into 3D. Here's our matrix and its column space:

$$C\left(\begin{bmatrix} 3 & 0 \\ 5 & 2 \\ 1 & 2 \end{bmatrix}\right) = \lambda_1 \begin{bmatrix} 3 \\ 5 \\ 1 \end{bmatrix} + \lambda_2 \begin{bmatrix} 0 \\ 2 \\ 2 \end{bmatrix}, \quad \lambda \in \mathbb{R}$$

Now there are two columns in $\mathbb{R}^3$. Those two columns are linearly independent, meaning you cannot express one as a scaled version of the other. So the column space of this matrix is 2D, but it's a 2D plane that is embedded in $\mathbb{R}^3$ (Figure 6-3).

The column space of this matrix is an infinite 2D plane, but that plane is merely an infinitesimal slice of 3D. You can think of it like an infinitely thin piece of paper that spans the universe.

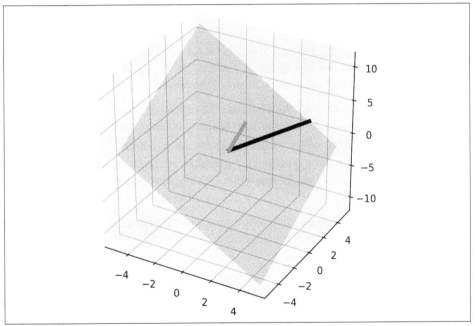

Figure 6-3. The 2D column space of a matrix embedded in 3D. The two thick lines depict the two columns of the matrix.

There are many vectors on that plane (i.e., many vectors that can be obtained as a linear combination of the two column vectors), but many *more* vectors that are not on the plane. In other words, there are vectors in the column space of the matrix, and there are vectors outside the column space of the matrix.

How do you know if a vector is in the column space of a matrix? That is not at all a trivial question—in fact, that question is the foundation of the linear least squares method, and words cannot do justice to the importance of least squares in applied math and engineering. So how *do* you figure out if a vector is in the column space? In the examples so far, we just used some guessing, arithmetic, and visualization. The point of that approach was to instill intuition, but those are obviously not scalable methods for higher dimensions and for more complicated problems.

Quantitative methods to determine whether a vector is in the column space of a matrix rely on the concept of matrix rank, which you'll learn about later in this chapter. Until then, focus on the intuition that the columns of a matrix carve out a vector subspace, which may include the entire M-dimensional space or may be some smaller-dimensional subspace; and that an important question is whether some other vector is inside that subspace (meaning that vector can be expressed as a linear weighted combination of the columns of the matrix).

Row Space

Once you understand the column space of a matrix, the row space is really easy to understand. In fact, the row space of a matrix is the exact same concept but we consider all possible weighted combinations of the rows instead of the columns.

Row space is indicated as $R(\mathbf{A})$. And because the transpose operation swaps rows and columns, you can also write that the row space of a matrix is the column space of the matrix transposed, in other words, $R(\mathbf{A}) = C(\mathbf{A}^\mathrm{T})$. There are a few differences between the row and column spaces of a matrix; for example, the row space (but not the column space) is invariant to row-reduction operations. But that is beyond the scope of this chapter.

Because the row space equals the column space of the matrix transpose, these two matrix spaces are identical for symmetric matrices.

Null Spaces

The null space is subtly but importantly different from the column space. The column space can be succintly summarized as the following equation:

$$\mathbf{A}\mathbf{x} = \mathbf{b}$$

This can be translated into English as "Can we find some set of coefficients in $\mathbf{x}$ such that the weighted combination of columns in $\mathbf{A}$ produce vector $\mathbf{b}$?" If the answer is yes, then $\mathbf{b} \in C(\mathbf{A})$ and vector $\mathbf{x}$ tells us how to weight the columns of $\mathbf{A}$ to get to $\mathbf{b}$.

The null space, in contrast, can be succinctly summarized as the following equation:

$$\mathbf{Ay} = \mathbf{0}$$

This can be translated into English as "Can we find some set of coefficients in $\mathbf{y}$ such that the weighted combination of columns in $\mathbf{A}$ produces the zeros vector $\mathbf{0}$?"

A moment's inspection will reveal an answer that works for every possible matrix $\mathbf{A}$: set $\mathbf{y} = \mathbf{0}$! Obviously, multiplying all the columns by 0s will sum to the zeros vector. But that is a trivial solution, and we exclude it. Therefore, the question becomes "Can we find a set of weights—not all of which are 0—that produces the zeros vector?" Any vector $\mathbf{y}$ that can satisfy this equation is in the null space of $\mathbf{A}$, which we write as $N(\mathbf{A})$.

Let's start with a simple example. Before reading the following text, see if you can find such a vector $\mathbf{y}$:

$$\begin{bmatrix} 1 & -1 \\ -2 & 2 \end{bmatrix}$$

Did you come up with a vector? Mine is [7.34, 7.34]. I'm willing to wager a Vegas-sized bet that you did not come up with the same vector. You might have come up with [1, 1] or perhaps [−1, −1]. Maybe [2, 2]?

I think you see where this is going—there is an infinite number of vectors $\mathbf{y}$ that satisfy $\mathbf{Ay} = \mathbf{0}$ for that specific matrix $\mathbf{A}$. And all of those vectors can be expressed as some scaled version of any of these choices. Thus, the null space of this matrix can be expressed as:

$$N(\mathbf{A}) = \lambda \begin{bmatrix} 1 \\ 1 \end{bmatrix}, \quad \lambda \in \mathbb{R}$$

Here's another example matrix. Again, try to find a set of coefficients such that the weighted sum of the columns produces the zeros vector (that is, find $\mathbf{y}$ in $\mathbf{Ay} = \mathbf{0}$):

$$\begin{bmatrix} 1 & -1 \\ -2 & 3 \end{bmatrix}$$

I'm willing to place an even larger bet that you couldn't find such a vector. But it's not because I don't believe you could (I have a very high opinion of my readers!); it's because this matrix has no null space. Formally, we say that the null space of this matrix is the empty set: $N(\mathbf{A}) = \{\}$.

Look back at the two example matrices in this subsection. You'll notice that the first matrix contains columns that can be formed as scaled versions of other columns, whereas the second matrix contains columns that form an independent set. That's no coincidence: there is a tight relationship between the dimensionality of the null space and the linear independence of the columns in a matrix. The exact nature of that relationship is given by the rank-nullity theorem, which you'll learn about in the next chapter. But they key point is this: the null space is empty when the columns of the matrix form a linearly independent set.

At the risk of redundancy, I will repeat this important point: full-rank and full column-rank matrices have empty null spaces, whereas reduced-rank matrices have nonempty (nontrivial) null spaces.

The Python SciPy library contains a function to compute the null space of a matrix. Let's confirm our results using code:

```
A = np.array([ [1,-1],[-2,2] ])
B = np.array([ [1,-1],[-2,3] ])

print( scipy.linalg.null_space(A) )
print( scipy.linalg.null_space(B) )
```

Here's the output:

```
[[0.70710678]
 [0.70710678]]

[]
```

The second output ([]) is the empty set. Why did Python choose 0.70710678 as numerical values for the null space of A? Wouldn't it be easier to read if Python chose 1? Given the infinity of possible vectors, Python returned a *unit vector* (you can mentally compute the norm of that vector knowing that $\sqrt{1/2} \approx .7071$). Unit vectors are convenient to work with and have several nice properties, including numerical stability. Therefore, computer algorithms will often return unit vectors as bases for subspaces. You'll see this again with eigenvectors and singular vectors.

What does the null space look like? Figure 6-4 shows the row vectors and null space for matrix A.

Why did I plot the row vectors instead of the column vectors? It turns out that the row space is orthogonal to the null space. That's not for some bizarre esoteric reason; instead, it's written right into the definition of the null space as $\mathbf{Ay} = \mathbf{0}$. Rewriting that equation for each row of the matrix (a_i) leads to the expression $a_i\mathbf{y} = \mathbf{0}$; in other words, the dot product between each row and the null space vector is 0.

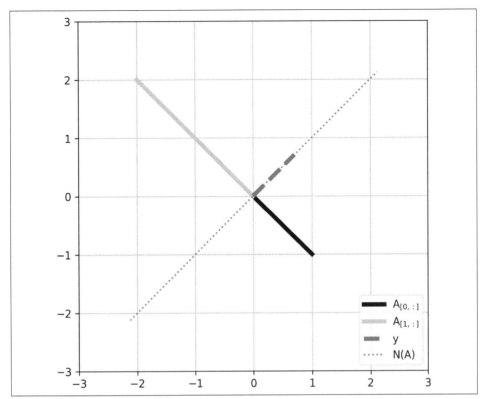

Figure 6-4. Visualization of the null space of a matrix

Why all the fuss about null spaces? It may seem odd to care about vectors that can multiply a matrix to produce the zeros vector. But the null space is the keystone of finding eigenvectors and singular vectors, as you will learn in Chapter 13.

Final thought for this section: every matrix has four associated subspaces; you've learned about three (column, row, null). The fourth subspace is called the *right null space*, and is the null space of the rows. It's often written as the null space of the matrix transpose: $N(\mathbf{A}^\mathrm{T})$. A traditional mathematics curriculum would now spend several weeks exploring the intricacies and relationships across the four subspaces. Matrix subspaces are worth studying for their fascinating beauty and perfection, but that's a level of depth to which we will not plunge.

Rank

Rank is a number associated with a matrix. It is related to the dimensionalities of matrix subspaces, and has important implications for matrix operations, including inverting matrices and determining the number of solutions to a system of equations.

As with other topics in this book, there are rich and detailed theories of matrix rank, but here I will focus on what you need to know for data science and related applications.

I will begin by listing a few properties of rank. In no specific order of importance:

- Rank is a nonnegative integer, so a matrix can have a rank of 0, 1, 2, …, but not −2 or 3.14.

- Every matrix has one unique rank; a matrix cannot simultaneously have multiple distinct ranks. (This also means that rank is a feature of the matrix, not of the rows or the columns.)

- The rank of a matrix is indicated using $r(\mathbf{A})$ or $rank(\mathbf{A})$. Also appropriate is "$\mathbf{A}$ is a rank-r matrix."

- The maximum possible rank of a matrix is the smaller of its row or column count. In other words, the maximum possible rank is min{M,N}.

- A matrix with its maximum possible rank is called "full-rank." A matrix with rank $r < $ min{M,N} is variously called "reduced-rank," "rank-deficient," or "singular."

- Scalar multiplication does not affect the matrix rank (with the exception of 0, which transforms the matrix into the zeros matrix with a rank of 0).

There are several equivalent interpretations and definitions of matrix rank. Those include:

- The largest number of columns (or rows) that form a linearly independent set.

- The dimensionality of the column space (which is the same as the dimensionality of the row space).

- The number of dimensions containing information in the matrix. This is not the same as the total number of columns or rows in the matrix, because of possible linear dependencies.

- The number of nonzero singular values of the matrix.

It may seem surprising that the definition of rank is the same for the columns and the rows: is the dimensionality really the same for the column space and the row space, even for a nonsquare matrix? Yes it is. There are various proofs of this, many of which are either fairly involved or rely on the singular value decomposition, so I won't include a formal proof in this chapter.[2] But I will show an example of the row and column spaces of a nonsquare matrix as an illustration.

Here is our matrix:

$$\begin{bmatrix} 1 & 1 & -4 \\ 2 & -2 & 2 \end{bmatrix}$$

The column space of the matrix is in $\mathbb{R}^2$ while the row space is in $\mathbb{R}^3$, so those two spaces need to be drawn in different graphs (Figure 6-5). The three columns do not form a linearly independent set (any one column can be described as a linear combination of the other two), but they do span all of $\mathbb{R}^2$. Therefore, the column space of the matrix is 2D. The two rows do form a linearly independent set, and the subspace they span is a 2D plane in $\mathbb{R}^3$.

To be clear: the column space and row space of the matrix are *different*. But the *dimensionality* of those matrix spaces is the same. And that dimensionality is the rank of the matrix. So this matrix has a rank of 2.

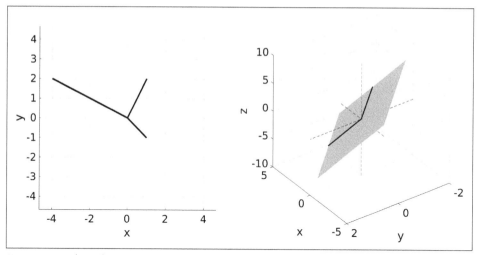

Figure 6-5. The column and row spaces have different spans but the same dimensionalities

---

2 If you are already familiar with the SVD, then the short version is that the SVD of $\mathbf{A}$ and $\mathbf{A}^T$ swap row and column spaces, but the number of nonzero singular values remains the same.

Following are some matrices. Although I haven't yet taught you how to compute rank, try to guess the rank of each matrix based on the previous descriptions. The answers are in the footnote.[3]

$$A = \begin{bmatrix} 1 \\ 2 \\ 4 \end{bmatrix}, \quad B = \begin{bmatrix} 1 & 3 \\ 2 & 6 \\ 4 & 12 \end{bmatrix}, \quad C = \begin{bmatrix} 1 & 3.1 \\ 2 & 6 \\ 4 & 12 \end{bmatrix}, \quad D = \begin{bmatrix} 1 & 3 & 2 \\ 6 & 6 & 1 \\ 4 & 2 & 0 \end{bmatrix}, \quad E = \begin{bmatrix} 1 & 1 & 1 \\ 1 & 1 & 1 \\ 1 & 1 & 1 \end{bmatrix}, \quad F = \begin{bmatrix} 0 & 0 & 0 \\ 0 & 0 & 0 \\ 0 & 0 & 0 \end{bmatrix}$$

I hope you managed to figure out the ranks, or at least weren't shocked to see the answers in the footnote.

Needless to say, visual inspection and some intuition is not a scalable method for computing rank in practice. There are several methods to compute rank. For example, in Chapter 10 you'll learn that you can compute the rank by row reducing a matrix to its echelon form and counting the number of pivots. Computer programs like Python compute rank by counting the number of nonzero singular values of the matrix. You'll learn about that in Chapter 14.

For now, I would like you to focus on the idea that the rank corresponds to the largest number of columns that can form a linearly independent set, which also corresponds to the dimensionality of the column space of the matrix. (You can replace "column" with "row" in the previous sentence and it will be equally accurate.)

Ranks of Special Matrices

Some special matrices have ranks that are easy to compute or that are worth learning about:

Vectors
>   All vectors have a rank of 1. That's because vectors—by definition—have only one column (or row) of information; the subspace they span is 1D. The only exception is the zeros vector.

Zeros matrices
>   The zeros matrix of any size (including the zeros vector) has a rank of 0.

Identity matrices
>   The rank of the identity matrix equals the number of rows (which equals the number of columns). In other words, $r(I_N) = N$. In fact, the identity matrix is simply a special case of a diagonal matrix.

---

3 $r(A) = 1$, $r(B) = 1$, $r(C) = 2$, $r(D) = 3$, $r(E) = 1$, $r(F) = 0$

Diagonal matrices

The rank of a diagonal matrix equals the number of nonzero diagonal elements. This is because each row contains maximum one nonzero element, and it is impossible to create a nonzero number through weighted combinations of zeros. This property becomes useful when solving systems of equations and interpreting the singular value decomposition.

Triangular matrices

A triangular matrix is full rank only if there are nonzero values in all diagonal elements. A triangular matrix with at least one zero in the diagonal will be reduced rank (the exact rank will depend on the numerical values in the matrix).

Random matrices

The rank of a random matrix is impossible to know a priori, because it depends on the distribution of numbers from which the elements in the matrix were drawn and on the probability of drawing each number. For example, a 2×2 matrix populated with either 0s or 1s could have a rank of 0 if the individual elements all equal 0. Or it could have a rank of 2, e.g., if the identity matrix is randomly selected.

But there is a way to create random matrices with guaranteed maximum possible rank. This is done by drawing floating-point numbers at random, e.g., from a Gaussian or uniform distribution. A 64-bit computer can represent 2^{64} numbers. Drawing a few dozen or hundred from that set to put in a matrix means that the likelihood of linear dependencies in the columns of the matrix is astronomically unlikely. So unlikely, in fact, that it could power the Infinite Improbability Drive of the *Heart of Gold*.[4]

The point is that matrices created via, e.g., `np.random.randn()` will have the maximum possible rank. This is useful for using Python to learn linear algebra, because you can create a matrix with any arbitrary rank (subject to constraints discussed earlier). Exercise 6-5 will guide you through the process.

Rank-1 matrices

A rank-1 matrix has—you guessed it—a rank of 1. This means that there is actually only one column's worth of information (alternatively: there is only one row's worth of information) in the matrix, and all other columns (or rows) are simply linear multiples. You saw a few examples of rank-1 matrices earlier in this chapter. Here are a few more:

---

4 The Infinite Improbability Drive is a technology that allows the *Heart of Gold* spaceship to traverse impossible distances in space. If you don't know what I'm talking about, then you haven't read *Hitchhiker's Guide to the Galaxy* by Douglas Adams. And if you haven't read that book, then you're really missing out on one of the greatest, most thought-provoking, and funniest intellectual achievements of the 20th century.

$$\begin{bmatrix} -2 & -4 & -4 \\ -1 & -2 & -2 \\ 0 & 0 & 0 \end{bmatrix}, \quad \begin{bmatrix} 2 & 1 \\ 0 & 0 \\ 2 & 1 \\ 4 & 2 \end{bmatrix}, \quad \begin{bmatrix} 12 & 4 & 4 & 12 & 4 \\ 6 & 2 & 2 & 6 & 2 \\ 9 & 3 & 3 & 9 & 3 \end{bmatrix}$$

Rank-1 matrices can be square, tall, or wide; regardless of the size, each column is a scaled copy of the first column (or each row is a scaled copy of the first row).

How does one create a rank-1 matrix? Actually, you already learned how in Chapter 2 (although I only briefly mentioned it; you are forgiven for forgetting). The answer is by taking the outer product between two nonzeros vectors. For example, the third matrix above is the outer product of $[4\ 2\ 3]^T$ and $[3\ 1\ 1\ 3\ 1]$.

Rank-1 matrices are important in eigendecomposition and the singular value decomposition. You'll encounter many rank-1 matrices in later chapters in this book—and in your adventures in applied linear algebra.

Rank of Added and Multiplied Matrices

If you know the ranks of matrices $\mathbf{A}$ and $\mathbf{B}$, do you automatically know the rank of $\mathbf{A} + \mathbf{B}$ or $\mathbf{AB}$?

The answer is no, you don't. But, the ranks of the two individual matrices provide upper bounds for the maximum possible rank. Here are the rules:

$$rank(\mathbf{A} + \mathbf{B}) \leq rank(\mathbf{A}) + rank(\mathbf{B})$$

$$rank(\mathbf{AB}) \leq \min \{rank(\mathbf{A}), rank(\mathbf{B})\}$$

I don't recommend memorizing these rules. But I do recommend memorizing the following:

- You cannot know the exact rank of a summed or product matrix based on knowing the ranks of the individual matrices (with a few exceptions, e.g., the zeros matrix); instead, the individual matrices provide upper bounds for the rank of the summed or product matrix.

- The rank of a summed matrix could be greater than the ranks of the individual matrices.

- The rank of a product matrix cannot be greater than the largest rank of the multiplying matrices.[5]

In Exercise 6-6, you will have the opportunity to illustrate these two rules using random matrices of various ranks.

Rank of Shifted Matrices

Simply put: shifted matrices have full rank. In fact, one of the primary goals of shifting a square matrix is to increase its rank from $r < M$ to $r = M$.

An obvious example is shifting the zeros matrix by the identity matrix. The rank of the resulting sum $\mathbf{0} + \mathbf{I}$ is a full-rank matrix.

Here's another, slightly less obvious, example:

$$\begin{bmatrix} 1 & 3 & 2 \\ 5 & 7 & 2 \\ 2 & 2 & 0 \end{bmatrix} + .01 \begin{bmatrix} 1 & 0 & 0 \\ 0 & 1 & 0 \\ 0 & 0 & 1 \end{bmatrix} = \begin{bmatrix} 1.01 & 3 & 2 \\ 5 & 7.01 & 2 \\ 2 & 2 & .01 \end{bmatrix}$$

The leftmost matrix has a rank of 2; notice that the third column equals the second minus the first. But the rank of the summed matrix is 3: the third column can no longer be produced by some linear combination of the first two. And yet, the information in the matrix has hardly changed; indeed, the Pearson correlation between the elements in the original and shifted matrix is $\rho = 0.999997222233796$. This has significant implications: for example, the rank-2 matrix cannot be inverted whereas the shifted matrix can. (You'll learn why in Chapter 8.)

---

5 This rule explains why the outer product always produces a rank-1 matrix.

Theory and Practice

Understanding Singular Values Before Understanding Singular Values

It is impossible to learn mathematics in a purely monotonic fashion, meaning you fully understand concept a before fully learning about concept b and so on. A complete understanding of matrix rank requires knowing about the singular value decomposition (SVD), but the SVD doesn't make sense until you know about rank. It's a bit of a catch-22. This is part of what makes math frustrating to learn. The good news is that pages in this book continue to exist after you read them, so if the following discussion isn't completely sensible, come back after learning about the SVD.

Briefly: every $M \times N$ matrix has a set of min$\{M,N\}$ nonnegative singular values that encode the "importance" or "spaciousness" of different directions in the column and row spaces of the matrix. Directions with a singular value of 0 are in one of the null spaces.

In abstract linear algebra, the rank is a rock-solid concept. Each matrix has exactly one rank, and that's the end of the story.

In practice, however, computing the matrix rank entails some uncertainty. Arguably, computers don't even *compute* the rank; they *estimate* it to a reasonable degree of accuracy. I wrote earlier that rank can be computed as the number of nonzero singular values. But that's not what Python does. Following are two key lines excerpted from np.linalg.matrix_rank() (I've dropped a few arguments to focus on the main point):

```
S = svd(M)
return count_nonzero(S > tol)
```

M is the matrix in question, S is a vector of singular values, and tol is a tolerance threshold. This means that NumPy is not actually counting *nonzero* singular values; it is counting singular values that are larger than some threshold. The exact threshold value depends on the numerical values in the matrix, but is generally around 10 to 12 orders of magnitude smaller than the matrix elements.

This means that NumPy makes a decision about what numbers are small enough to be considered "effectively zero." I am certainly not criticizing NumPy—this is the right thing to do! (Other numerical processing programs like MATLAB and Julia compute rank in the same way.)

But why do this? Why not simply count the nonzero singular values? The answer is that the tolerance absorbs small numerical inaccuracies that may arise due to computer rounding error. Having a tolerance also allows for ignoring tiny amounts of noise that might, for example, contaminate the data-acquisition sensors. This idea is used in data cleaning, compression, and dimension reduction. Figure 6-6 illustrates the concept.

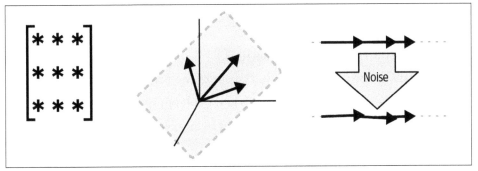

Figure 6-6. A 3 × 3 matrix representing a 2D plane may be considered rank-3 in the presence of a small amount of noise. The rightmost diagram shows the perspective of looking directly across the surface of the plane.

Rank Applications

There are many applications of matrix rank. In this section, I will introduce two of them.

In the Column Space?

In Chapter 4, you learned about the column space of a matrix, and you learned that an important question in linear algebra is whether a vector is in the column space of a matrix (which can be mathematically written as $\mathbf{v} \in C(\mathbf{A})$?). I also wrote that answering the question in a rigorous and scalable way depends on understanding matrix rank.

Before telling you the algorithm for determining whether $\mathbf{v} \in C(\mathbf{A})$, I need to briefly introduce a procedure called *augmenting* a matrix.

To augment a matrix means to add extra columns to the right-hand side of the matrix. You start with the "base" $M \times N$ matrix and the "extra" $M \times K$ matrix. The augmented matrix is of size $M \times (N + K)$. Augmenting two matrices is valid as long as they have the same number of rows (they can have a different number of columns). You'll see augmented matrices here in this section, and again in Chapter 10 when solving systems of equations.

Following is an example that illustrates the procedure:

$$\begin{bmatrix} 4 & 5 & 6 \\ 0 & 1 & 2 \\ 9 & 9 & 4 \end{bmatrix} \sqcup \begin{bmatrix} 1 \\ 2 \\ 3 \end{bmatrix} = \begin{bmatrix} 4 & 5 & 6 & 1 \\ 0 & 1 & 2 & 2 \\ 9 & 9 & 4 & 3 \end{bmatrix}$$

With that prerequisite tangent out the way, here is the algorithm for determining whether a vector is in the column space of a matrix:

1. **Augment the matrix with the vector.** Let's call the original matrix **A** and the augmented matrix $\widetilde{\mathbf{A}}$.
2. **Compute the ranks of the two matrices.**
3. **Compare the two ranks.** There will be one of two possible outcomes:
   a. $rank(\mathbf{A}) = rank(\widetilde{\mathbf{A}})$ Vector **v** is in the column space of matrix **A**.
   b. $rank(\mathbf{A}) < rank(\widetilde{\mathbf{A}})$ Vector **v** is **not** in the column space of matrix **A**.

What is the logic behind this test? If $\mathbf{v} \in C(\mathbf{A})$, then **v** can be expressed as some linear weighted combination of the columns of **A** (the columns of the augmented matrix $\widetilde{\mathbf{A}}$ form a linearly dependent set). In terms of span, vector **v** is redundant in $\widetilde{\mathbf{A}}$. Hence, the rank stays the same.

Conversely, if $\mathbf{v} \notin C(\mathbf{A})$, then **v** cannot be expressed as a linear weighted combination of the columns of **A**, which means that **v** has added new information into $\widetilde{\mathbf{A}}$. And that means that the rank will increase by 1.

Determining whether a vector is in the column space of a matrix is not merely some esoteric academic exercise; it is part of the reasoning behind linear least squares modeling, which is the math that underlies ANOVAs, regressions, and general linear models. You'll learn more about this in Chapters 10 and 11, but the basic idea is that we develop a model of how the world works and then translate that model into a matrix, which is called the *design matrix*. The data we measure from the world are stored in a vector. If that data vector is in the column space of the design matrix, then we have perfectly modeled the world. In nearly all cases, the data vector is not in the column space, and therefore we determine whether it is close enough to be considered statistically significant. Obviously, I will go into much greater detail in later chapters, but I'm foreshadowing here to keep up your enthusiasm for learning.

Linear Independence of a Vector Set

Now that you know about matrix rank, you are ready to understand an algorithm for determining whether a set of vectors is linearly independent. If you'd like to come up with such an algorithm on your own, then feel free to do so now before reading the rest of this section.

The algorithm is straightforward: put the vectors into a matrix, compute the rank of the matrix, and then compare that rank to the maximum possible rank of that matrix (remember that this is min{M,N}; for convenience of discussion I will assume that we have a tall matrix). The possible outcomes are:

- $r = M$: The vector set is linearly independent.

- $r < M$: The vector set is linearly dependent.

The reasoning behind this algorithm should be clear: if the rank is smaller than the number of columns, then at least one column can be described as a linear combination of other columns, which is the definition of linear dependence. If the rank equals the number of columns, then each column contributes unique information to the matrix, which means that no column can be described as a linear combination of other columns.

There is a more general point I'd like to make here: many operations and applications in linear algebra are actually quite simple and sensible, but they require a significant amount of background knowledge to understand. That's a good thing: the more linear algebra you know, the easier linear algebra becomes.

On the other hand, that statement is not universally true: there are operations in linear algebra that are so mind-numbingly tedious and complicated that I cannot bring myself to describe them in full detail in this book. Segue to the next section…

Determinant

The *determinant* is a number associated with a square matrix. In abstract linear algebra, the determinant is a keystone quantity in several operations, including the matrix inverse. But computing the determinant in practice can be numerically unstable for large matrices, due to underflow and overflow issues. You'll see an illustration of this in the exercises.

Nonetheless, you cannot understand the matrix inverse or eigendecomposition without understanding the determinant.

The two most important properties of the determinant—and the two most important take-home messages of this section—are (1) it is defined only for square matrices and (2) it is zero for singular (reduced-rank) matrices.

The determinant is notated as $det(\mathbf{A})$ or $|\mathbf{A}|$ (note the single vertical lines instead of the double vertical lines that indicate the matrix norm). The Greek capital delta Δ is used when you don't need to refer to a specific matrix.

But what *is* the determinant? What does it mean, and how do we interpret it? The determinant has a geometric interpretation, which is related to how much the matrix stretches vectors during matrix-vector multiplication (recall from the previous chapter that matrix-vector multiplication is a mechanism of applying geometric transforms to a coordinate expressed as a vector). A negative determinant means that one axis is rotated during the transformation.

However, in data-science-related applications, the determinant is used algebraically; these geometric explanations are not very insightful for how we use the determinant to find eigenvalues or invert a data covariance matrix. Therefore, suffice it to say that the determinant is a crucial step in advanced topics like the matrix inverse, eigendecomposition, and singular value decomposition, and that you might save yourself some stress and sleepless nights by simply accepting that the determinant is a tool in our toolbox without worrying so much about what it means.

Computing the Determinant

Computing the determinant is time-consuming and tedious. If I live a thousand years and never hand-calculate the determinant of a 5×5 matrix, I will have lived a rich, full, meaningful life. That said, there is a shortcut for computing the determinant of a 2×2 matrix, which is shown in Equation 6-2.

Equation 6-2. Computing the determinant of a 2×2 matrix

$$det\left(\begin{bmatrix} a & b \\ c & d \end{bmatrix}\right) = \begin{vmatrix} a & b \\ c & d \end{vmatrix} = ad - bc$$

You can see from this equation that the determinant is not limited to integers or positive values. Depending on the numerical values in the matrix, the determinant can be -1223729358 or $+.00000002$ or any other number. (For a real-valued matrix, the determinant will always be a real number.)

Calculating the determinant is really simple, right? It's just the product of the diagonal minus the product of the off-diagonal. For many matrices, you can compute the determinant in your head. The determinant is even simpler for a 1×1 matrix: it's simply the absolute value of that number.

You might now be doubting my claim that the determinant is numerically unstable.

The thing is that the shortcut for computing the determinant of a 2×2 matrix doesn't scale up to larger matrices. There is a "shortcut" for 3×3 matrices, but it isn't really a shortcut; it's a visual mnemonic. I won't show it here, but I will write out the upshot:

$$\begin{vmatrix} a & b & c \\ d & e & f \\ g & h & i \end{vmatrix} = aei + bfg + cdh - ceg - bdi - afh$$

Once you get to 4×4 matrices, the determinant becomes a real hassle to compute, unless the matrix has many carefully placed zeros. But I know you're curious, so Equation 6-3 shows the determinant of a 4×4 matrix.

Equation 6-3. Yeah, good luck with that

$$\begin{vmatrix} a & b & c & d \\ e & f & g & h \\ i & j & k & l \\ m & n & o & p \end{vmatrix} = \begin{array}{l} afkp - aflo - agjp + agln + ahjo - ahkn - bekp + belo \\ + bgip - bglm - bhio + bhkm + cejp - celn - cfip + cflm \\ + chin - chjm - dejo + dekn + dfio - dfkm - dgin + dgjm \end{array}$$

I'm not even going to show the full procedure for computing the determinant of any size matrix, because, let's be honest: you are reading this book because you're interested in applied linear algebra; it's important to understand how to use the determinant, not the full formula for computing it.

Anyway, the point is that if you ever need to compute the determinant, you use `np.linalg.det()` or `scipy.linalg.det()`.

Determinant with Linear Dependencies

The second take-home message about determinants is that they are zero for any reduced-rank matrix. We can explore this with a 2×2 matrix. Remember that any reduced-rank matrix has at least one column that can be expressed as a linear combination of other columns:

$$\begin{vmatrix} a & \lambda a \\ c & \lambda c \end{vmatrix} = ac\lambda - a\lambda c = 0$$

Here's the determinant of a 3×3 singular matrix:

$$\begin{vmatrix} a & b & \lambda a \\ d & e & \lambda d \\ g & h & \lambda g \end{vmatrix} = ae\lambda g + b\lambda dg + \lambda adh - \lambda aeg - bd\lambda g - a\lambda dh = 0$$

This concept generalizes to larger matrices. Thus, all reduced-rank matrices have a determinant of 0. The actual rank doesn't matter; if $r < M$, then $\Delta = 0$. All full-rank matrices have a nonzero determinant.

I've already written that I think geometric interpretations of the determinant have limited value to understanding why the determinant is important in data science. But $\Delta = 0$ does have a nice geometric meaning: a matrix with $\Delta = 0$ is a transformation in which at least one dimension gets flattened to have surface area but no volume.

You can imagine squashing a ball down to an infinitely thin disk. You will see a visual example of this in the next chapter ("Geometric Transformations via Matrix-Vector Multiplication" on page 116).

The Characteristic Polynomial

The equation for the determinant of a 2×2 matrix has five quantities: the four elements in the matrix and the determinant value. We can write them out in an equation as $ad - bc = \Delta$. One of the great things about equations is that you can move quantities around and solve for different variables. Consider Equation 6-4; assume that a, b, c, and Δ are known, and λ is some unknown quantity.

Equation 6-4. Using the determinant to find a missing matrix element

$$\begin{vmatrix} a & b \\ c & \lambda \end{vmatrix} \quad \Rightarrow \quad a\lambda - bc = \Delta$$

A bit of middle-school algebra will enable you to solve for λ in terms of the other quantities. The solution itself is unimportant; the point is that *if we know the determinant of a matrix, we can solve for unknown variables inside the matrix.*

Here's a numerical example:

$$\begin{vmatrix} 2 & 7 \\ 4 & \lambda \end{vmatrix} = 4 \quad \Rightarrow \quad 2\lambda - 28 = 4 \quad \Rightarrow \quad 2\lambda = 32 \quad \Rightarrow \quad \lambda = 16$$

Now let's take this one step further:

$$\begin{vmatrix} \lambda & 1 \\ 3 & \lambda \end{vmatrix} = 1 \quad \Rightarrow \quad \lambda^2 - 3 = 1 \quad \Rightarrow \quad \lambda^2 = 4 \quad \Rightarrow \quad \lambda = \pm 2$$

The same unknown λ was on the diagonal, which produced a second-order polynomial equation, and that produced two solutions. That's no coincidence: the fundamental theorem of algebra tells us that an nth order polynomial has exactly n roots (though some may be complex valued).

Here is the key point that I'm working towards: combining matrix shifting with the determinant is called the *characteristic polynomial* of the matrix, as shown in Equation 6-5.

Equation 6-5. The characteristic polynomial of the matrix

$$det(\mathbf{A} - \lambda\mathbf{I}) = \Delta$$

Why is it called a polynomial? Because the shifted $M \times M$ matrix has an λ^M term, and therefore has M solutions. Here's what it looks like for 2×2 and 3×3 matrices:

$$\begin{vmatrix} a - \lambda & b \\ c & d - \lambda \end{vmatrix} = 0 \quad \Rightarrow \quad \lambda^2 - (a + d)\lambda + (ad - bc) = 0$$

$$\begin{vmatrix} a - \lambda & b & c \\ d & e - \lambda & f \\ g & h & i - \lambda \end{vmatrix} = 0 \quad \Rightarrow \quad \begin{matrix} -\lambda^3 + (a + e + i)\lambda^2 \\ +(-ae + bd - ai + cg - ei + fh)\lambda \\ +aei - afh - bdi + bfg + cdh - ceg = 0 \end{matrix}$$

Please don't ask me to show you the characteristic polynomial for a 4×4 matrix. Trust me, it will have a λ^4 term.

Let's return to the 2×2 case, this time using numbers instead of letters. The matrix below is full rank, meaning that its determinant is not 0 (it is $\Delta = -8$), but I'm going to assume that it has a determinant of 0 after being shifted by some scalar λ; the question is, what values of λ will make this matrix reduced-rank? Let's use the characteristic polynomial to find out:

$$det\left(\begin{bmatrix} 1 & 3 \\ 3 & 1 \end{bmatrix} - \lambda \mathbf{I}\right) = 0$$

$$\begin{vmatrix} 1 - \lambda & 3 \\ 3 & 1 - \lambda \end{vmatrix} = 0 \quad \Rightarrow \quad (1 - \lambda)^2 - 9 = 0$$

After some algebra (which I encourage you to work through), the two solutions are $\lambda = -2, 4$. What do these numbers mean? To find out, let's plug them back into the shifted matrix:

$$\lambda = -2 \quad \Rightarrow \quad \begin{bmatrix} 3 & 3 \\ 3 & 3 \end{bmatrix}$$

$$\lambda = 4 \quad \Rightarrow \quad \begin{bmatrix} -3 & 3 \\ 3 & -3 \end{bmatrix}$$

Clearly, both matrices are rank-1. Furthermore, both matrices have nontrivial null spaces, meaning you can find some nonzeros vector $\mathbf{y}$ such that $(\mathbf{A} - \lambda\mathbf{I})\mathbf{y} = \mathbf{0}$. I'm sure you can find a vector for each of those λs on your own! But just in case you want to confirm your correct answer, check the footnote.[6]

The characteristic polynomial is, to use a technical term, super awesome. For one thing, it provides the remarkable insight that every square *matrix* can be expressed as an *equation*. And not just any equation—an equation that directly links matrices to the fundamental theorem of algebra. And if that isn't cool enough, the solutions to the characteristic polynomial set to $\Delta = 0$ are the eigenvalues of the matrix (that's the λs we found above). I know that you have to wait until Chapter 13 to understand what eigenvalues are and why they are important, but you can sleep well tonight knowing that today you learned how to uncover the eigenvalues of a matrix.

Summary

The goal of this chapter was to expand your knowledge of matrices to include several important concepts: norms, spaces, rank, and determinant. If this were a book on abstract linear algebra, this chapter could easily have been a hundred-plus pages. But I tried to focus the discussion on what you'll need to know for linear algebra applications in data science and AI. You'll discover more about matrix spaces in later chapters (in particular, least squares fitting in Chapter 11 and the singular value decomposition in Chapter 14). In the meantime, here is a list of the most important points in this chapter:

- There are many kinds of matrix norms, which can be broadly classified into *element-wise* and *induced*. The former reflects the magnitudes of the elements in the matrix while the latter reflects the geometric-transformative effect of the matrix on vectors.

- The most commonly used element-wise norm is called the Frobenius norm (a.k.a. the Euclidean norm or the $\ell 2$ norm) and is computed as the square root of the sum of squared elements.

- The trace of a matrix is the sum of the diagonal elements.

- There are four matrix spaces (column, row, null, right-null), and they are defined as the set of linear weighted combinations of different features of the matrix.

- The column space of the matrix comprises all linear weighted combinations of the columns in the matrix and is written as $C(\mathbf{A})$.

- An important question in linear algebra is whether some vector $\mathbf{b}$ is in the column space of a matrix; if it is, then there is some vector $\mathbf{x}$ such that $\mathbf{Ax} = \mathbf{b}$. Answering that question forms the basis for many statistical model fitting algorithms.

---

6 Any scaled version of [1 −1] and [1 1].

- The row space of a matrix is the set of linear weighted combintions of the rows of the matrix and is indicated as $R(\mathbf{A})$ or $C(\mathbf{A}^{\mathrm{T}})$.

- The null space of a matrix is the set of vectors that linearly combines the columns to produce the zeros vector—in other words, any vector $\mathbf{y}$ that solves the equation $\mathbf{Ay} = \mathbf{0}$. The trivial solution of $\mathbf{y} = \mathbf{0}$ is excluded. The null space is important for finding eigenvectors, among other applications.

- Rank is a nonnegative integer associated with a matrix. Rank reflects the largest number of columns (or rows) that can form a linearly independent set. Matrices with a rank smaller than the maximum possible are called reduced-rank or singular.

- Shifting a square matrix by adding a constant to the diagonal ensures full-rank.

- One (of many) applications of rank is to determine whether a vector is in the column space of a matrix, which works by comparing the rank of the matrix to the rank of the vector-augmented matrix.

- The determinant is a number associated with a square matrix (there is no determinant for rectangular matrices). It is tedious to compute, but the most important thing to know about the determinant is that it is zero for all reduced-rank matrices and nonzero for all full-rank matrices.

- The characteristic polynomial transforms a square matrix, shifted by λ, into an equation that equals the determinant. Knowing the determinant allows you to solve for λ. I've only briefly hinted at why that is so important, but trust me: it's important. (Or don't trust me and see for yourself in Chapter 13.)

Code Exercises

Exercise 6-1.

The norm of a matrix is related to the scale of the numerical values in the matrix. In this exercise, you will create an experiment to demonstrate this. In each of 10 experiment iterations, create a 10×10 random numbers matrix and compute its Frobenius norm. Then repeat this experiment 40 times, each time scalar multiplying the matrix by a different scalar that ranges between 0 and 50. The result of the experiment will be a 40×10 matrix of norms. Figure 6-7 shows the resulting norms, averaged over the 10 experiment iterations. This experiment also illustrates two additional properties of matrix norms: they are strictly nonnegative and can equal 0 only for the zeros matrix.

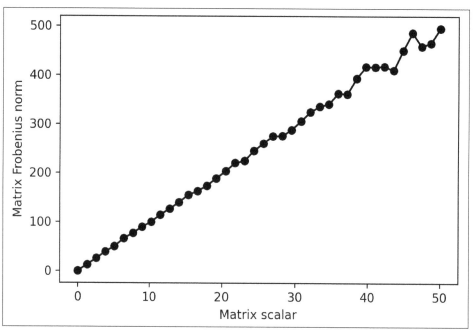

Figure 6-7. Solution to Exercise 6-1

Exercise 6-2.

In this exercise, you will write an algorithm that finds a scalar that brings the Frobenius distance between two matrices to 1. Start by writing a Python function that takes two matrices (of the same size) as input and returns the Frobenius distance between them. Then create two $N \times N$ random numbers matrices (I used $N = 7$ in the solutions code, but you can use any other size). Create a variable s = 1 that scalar multiplies both matrices. Compute the Frobenius distance between the scaled matrices. As long as that distance remains above 1, set the scalar to be .9 times itself and recompute the distance between the scaled matrices. This should be done in a while loop. When the Frobenius distance gets below 1, quit the while loop and report the number of iterations (which corresponds to the number of times that the scalar s was multiplied by .9) and the scalar value.

Exercise 6-3.

Demonstrate that the trace method and the Euclidean formula produce the same result (the Frobenius norm). Does the trace formula work only for $\mathbf{A}^\mathrm{T}\mathbf{A}$, or do you get the same result for $\mathbf{A}\mathbf{A}^\mathrm{T}$?

Exercise 6-4.

This will be a fun exercise,[7] because you'll get to incorporate material from this and the previous chapters. You will explore the impact of shifting a matrix on the norm of that matrix. Start by creating a 10×10 random matrix and compute its Frobenius norm. Then code the following steps inside a for loop: (1) shift the matrix by a fraction of the norm, (2) compute the percent change in norm from the original, (3) compute the Frobenius distance between the shifted and original matrices, and (4) compute the correlation coefficient between the elements in the matrices (hint: correlate the vectorized matrices using np.flatten()). The fraction of the norm that you shift by should range from 0 to 1 in 30 linearly spaced steps. Make sure that at each iteration of the loop, you use the original matrix, not the shifted matrix from the previous iteration. You should get a plot that looks like Figure 6-8.

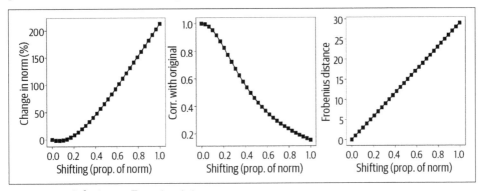

Figure 6-8. Solution to Exercise 6-4

Exercise 6-5.

I will now show you how to create random matrices with arbitrary rank (subject to the constraints about matrix sizes, etc.). To create an $M \times N$ matrix with rank r, multiply a random $M \times r$ matrix with an $r \times N$ matrix. Implement this in Python and confirm that the rank is indeed r. What happens if you set $r > \min\{M,N\}$, and why does that happen?

Exercise 6-6.

Demonstrate the addition rule of matrix rank ($r(\mathbf{A} + \mathbf{B}) \leq r(\mathbf{A}) + r(\mathbf{B})$) by creating three pairs of rank-1 matrices that have a sum with (1) rank-0, (2) rank-1, and (3) rank-2. Then repeat this exercise using matrix multiplication instead of addition.

---

7 *All* of these exercises are fun, but some more than others.

Exercise 6-7.

Put the code from Exercise 6-5 into a Python function that takes parameters M and r as input and provides a random $M \times M$ rank-r matrix as output. In a double for loop, create pairs of 20×20 matrices with individual ranks varying from 2 to 15. Add and multiply those matrices, and store the ranks of those resulting matrices. Those ranks can be organized into a matrix and visualized as a function of the ranks of the individual matrices (Figure 6-9).

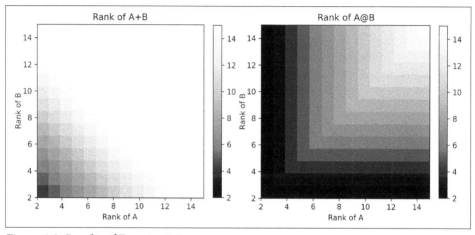

Figure 6-9. Results of Exercise 6-7

Exercise 6-8.

Interestingly, the matrices $\mathbf{A}$, $\mathbf{A}^T$, $\mathbf{A}^T\mathbf{A}$, and $\mathbf{A}\mathbf{A}^T$ all have the same rank. Write code to demonstrate this, using random matrices of various sizes, shapes (square, tall, wide), and ranks.

Exercise 6-9.

The goal of this exercise is to answer the question $\mathbf{v} \in C(\mathbf{A})$? Create a rank-3 matrix $\mathbf{A} \in \mathbb{R}^{4 \times 3}$ and vector $\mathbf{v} \in \mathbb{R}^4$ using numbers randomly drawn from a normal distribution. Follow the algorithm described earlier to determine whether the vector is in the column space of the matrix. Rerun the code multiple times to see whether you find a consistent pattern. Next, use a $\mathbf{A} \in \mathbb{R}^{4 \times 4}$ rank-4 matrix. I'm willing to bet one million bitcoins[8] that you *always* find that $\mathbf{v} \in C(\mathbf{A})$ when $\mathbf{A}$ is a 4×4 random matrix (assuming no coding mistakes). What makes me confident about your answer?[9]

For an extra challenge, put this code into a function that returns `True` or `False` depending on the outcome of the test, and that raises an exception (that is, a useful error message) if the size of the vector does not match for matrix augmentation.

Exercise 6-10.

Remember that the determinant of a reduced-rank matrix is—in theory—zero. In this exercise, you will put this theory to the test. Implement the following steps: (1) Create a square random matrix. (2) Reduce the rank of the matrix. Previously you've done this by multiplying rectangular matrices; here, set one column to be a multiple of another column. (3) Compute the determinant and store its absolute value.

Run these three steps in a double `for` loop: one loop over matrix sizes ranging from 3×3 to 30×30 and a second loop that repeats the three steps one hundred times (repeating an experiment is useful when simulating noise data). Finally, plot the determinant, averaged over the one hundred repeats, in a line plot as a function of the number of elements in the matrix. Linear algebra theory predicts that that line (that is, the determinants of all the reduced-rank matrices) is zero, regardless of the matrix size. Figure 6-10 shows otherwise, reflecting the computational difficulties with accurately computing the determinant. I log transformed the data for increased visibility; you should inspect the plot using log scaling and linear scaling.

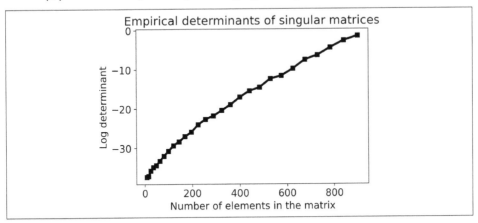

Figure 6-10. Results of Exercise 6-10

---

8 *n.b.* I don't have that many bitcoins, ¯_(ツ)_/¯.

9 Because the column space of a full-rank matrix spans all of $\mathbb{R}^M$, and therefore *all* vectors in $\mathbb{R}^M$ are necessarily in the column space.

## CHAPTER 7

Matrix Applications

I hope that now, after the past two theory-heavy chapters, you feel like you just finished an intense workout at the gym: exhausted but energized. This chapter should feel like a bike ride through the hills in the countryside: effortful at times but offering a fresh and inspiring perspective.

The applications in this chapter are loosely built off of those in Chapter 4. I did this to have some common threads that bind the chapters on vectors and on matrices. And because I want you to see that although the concepts and applications become more complex as you progress in linear algebra, the foundations are still built on the same simple principles such as linear weighted combinations and the dot product.

Multivariate Data Covariance Matrices

In Chapter 4, you learned how to compute the Pearson correlation coefficient as the vector dot product between two data variables, divided by the product of the vector norms. That formula was for two variables (e.g., height and weight); what if you have multiple variables (e.g., height, weight, age, weekly exercise…)?

You could imagine writing a double for loop over all of the variables, and applying the bivariate correlation formula to all pairs of variables. But that is cumbersome and inelegant, and therefore antithetical to the spirit of linear algebra. The purpose of this section is to show you how to compute covariance and correlation *matrices* from multivariate datasets.

Let's start with *covariance*. Covariance is simply the numerator of the correlation equation—in other words, the dot product between two mean-centered variables. Covariance is interpreted in the same way as correlation (positive when variables move together, negative when variables move apart, zero when the variables have no linear relationship), except that covariance retains the scale of the data, and therefore is not bound by ±1.

Covariance also has a normalization factor of $n - 1$, where n is the number of data points. That normalization prevents the covariance from growing larger as you sum more data values together (analogous to dividing by N to transform a sum into an average). Here is the equation for covariance:

$$c_{a,b} = (n-1)^{-1} \sum_{i=1}^{n} (x_i - \bar{x})(y_i - \bar{y})$$

As in Chapter 4, if we call $\tilde{\mathbf{x}}$ to be the mean-centered variable $\mathbf{x}$, then covariance is simply $\tilde{\mathbf{x}}^T \tilde{\mathbf{y}}/(n-1)$.

The key insight to implementing this formula for multiple variables is that matrix multiplication is an organized collection of dot products between rows of the left matrix and columns of the right matrix.

So here's what we do: create a matrix in which each column corresponds to each variable (a variable is a data feature). Let's call that matrix $\mathbf{X}$. Now, the multiplication $\mathbf{XX}$ is not sensible (and probably not even valid, because data matrices tend to be tall, thus $M > N$). But if we were to transpose the first matrix, then the *rows* of $\mathbf{X}^T$ correspond to the *columns* of $\mathbf{X}$. Therefore, the matrix product $\mathbf{X}^T\mathbf{X}$ encodes all of the pair-wise covariances (assuming the columns are mean centered, and when dividing by $n - 1$). In other words, the (i,j)th element in the covariance matrix is the dot product between data features i and j.

The matrix equation for a covariance matrix is elegant and compact:

$$\mathbf{C} = \mathbf{X}^T\mathbf{X}\frac{1}{n-1}$$

Matrix $\mathbf{C}$ is symmetric. That comes from the proof in Chapter 5 that any matrix times its transpose is square symmetric, but it also makes sense statistically: covariance and correlation are symmetric, meaning that, for example, the correlation between height and weight is the same as the correlation between weight and height.

What are the diagonal elements of $\mathbf{C}$? Those contain the covariances of each variable with itself, which in statistics is called the *variance*, and quantifies the dispersion around the mean (variance is the squared standard deviation).

Why Transpose the Left Matrix?

There is nothing special about transposing the left matrix. If your data matrix is organized as features-by-observations, then its covariance matrix is $\mathbf{XX}^T$. If you are ever uncertain about whether to transpose the left or right matrix, think about how to apply the rules for matrix multiplication to produce a features-by-features matrix. Covariance matrices are always features-by-features.

The example in the online code creates a covariance matrix from a publicly available dataset on crime statistics. The dataset includes over a hundred features about social, economic, educational, and housing information in various communities around the US.[1] The goal of the dataset is to use these features to predict levels of crime, but here we will use it to inspect covariance and correlation matrices.

After importing and some light data processing (explained in the online code), we have a data matrix called dataMat. The following code shows how to compute the covariance matrix:

```
datamean = np.mean(dataMat,axis=0) # vector of feature means
dataMatM = dataMat - datamean      # mean-center using broadcasting
covMat   = dataMatM.T @ dataMatM   # data matrix times its transpose
covMat   /= (dataMatM.shape[0]-1)  # divide by N-1
```

Figure 7-1 shows an image of the covariance matrix. First of all: it looks neat, doesn't it? I work with multivariate datasets in my "day job" as a neuroscience professor, and ogling covariance matrices has never failed to put a smile on my face.

In this matrix, light colors indicate variables that covary positively (e.g., percentage of divorced males versus number of people in poverty), dark colors indicate variables that covary negatively (e.g., percentage of divorced males versus median income), and gray colors indicate variables that are unrelated to each other.

As you learned in Chapter 4, computing a correlation from a covariance simply involves scaling by the norms of the vectors. This can be translated into a matrix equation, which will allow you to compute the data correlation matrix without for loops. Exercise 7-1 and Exercise 7-2 will walk you through the procedure. As I wrote in Chapter 4, I encourage you to work through those exercises before continuing to the next section.

---

1 M. A. Redmond and A. Baveja, "A Data-Driven Software Tool for Enabling Cooperative Information Sharing Among Police Departments," *European Journal of Operational Research* 141 (2002): 660–678.

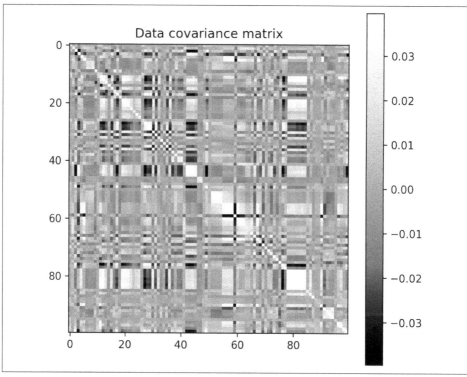

Figure 7-1. A data covariance matrix

Final note: NumPy has functions to compute covariance and correlation matrices (respectively, `np.cov()` and `np.corrcoef()`). In practice, it's more convenient to use those functions than to write out the code yourself. But—as always in this book—I want you to understand the math and mechanisms that those functions implement. Hence, for these exercises you should implement the covariance as a direct translation of the formulas instead of calling NumPy functions.

Geometric Transformations via Matrix-Vector Multiplication

I mentioned in Chapter 5 that one of the purposes of matrix-vector multiplication is to apply a geometric transform to a set of coordinates. In this section, you will see this in 2D static images and in animations. Along the way, you'll learn about pure rotation matrices and about creating data animations in Python.

A "pure rotation matrix" rotates a vector while preserving its length. You can think about the hands of an analog clock: as time ticks by, the hands rotate but do not change in length. A 2D rotation matrix can be expressed as:

$$\mathbf{T} = \begin{bmatrix} \cos{(\theta)} & \sin{(\theta)} \\ -\sin{(\theta)} & \cos{(\theta)} \end{bmatrix}$$

A pure rotation matrix is an example of an *orthogonal matrix*. I will write more about orthogonal matrices in the next chapter, but I would like to point out that the columns of **T** are orthogonal (their dot product is $\cos(\theta)\sin(\theta) - \sin(\theta)\cos(\theta)$) and are unit vectors (recall the trig identity that $\cos^2(\theta) + \sin^2(\theta) = 1$.)

To use this transformation matrix, set θ to some angle of clockwise rotation, and then muliply matrix **T** by a $2 \times N$ matrix of geometric points, where each column in that matrix contains the (X,Y) coordinates for each of N data points. For example, setting $\theta = 0$ does not change the points' locations (this is because $\theta = 0$ means $\mathbf{T} = \mathbf{I}$); setting $\theta = \pi/2$ rotates the points by 90° around the origin.

As a simple example, consider a set of points aligned in a vertical line and the effect of multiplying those coordinates by **T**. In Figure 7-2, I set $\theta = \pi/5$.

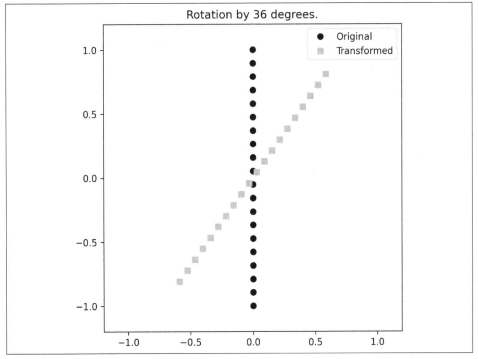

Figure 7-2. Twirling points around the origin through a pure rotation matrix

Before continuing with this section, please inspect the online code that generates this figure. Make sure you understand how the math I wrote above is translated into code, and take a moment to experiment with different rotation angles. You can also try to figure out how to make the rotation counterclockwise instead of clockwise; the answer is in the footnote.[2]

Let's make our investigations of rotations more exciting by using "impure" rotations (that is, stretching and rotating, not only rotating) and by animating the transformations. In particular, we will smoothly adjust the transformation matrix at each frame of a movie.

There are several ways to create animations in Python; the method I'll use here involves defining a Python function that creates the figure content on each movie frame, then calling a `matplotlib` routine to run that function on each iteration of the movie.

I call this movie *The Wobbly Circle*. Circles are defined by a set of $\cos(\theta)$ and $\sin(\theta)$ points, for a vector of θ angles that range from 0 to 2π.

I set the transformation matrix to the following:

$$\mathbf{T} = \begin{bmatrix} 1 & 1-\phi \\ 0 & 1 \end{bmatrix}$$

Why did I pick these specific values, and how do you interpret a transformation matrix? In general, the diagonal elements scale the *x*-axis and *y*-axis coordinates, while the off-diagonal elements stretch both axes. The exact values in the matrix above were selected by playing around with the numbers until I found something that I thought looked neat. Later on, and in the exercises, you will have the opportunity to explore the effects of changing the transformation matrix.

Over the course of the movie, the value of ϕ will smoothly transition from 1 to 0 and back to 1, following the formula $\phi = x^2$, $-1 \le x \le 1$. Note that $\mathbf{T} = \mathbf{I}$ when $\phi = 1$.

The Python code for a data animation can be separated into three parts. The first part is to set up the figure:

```
theta  = np.linspace(0,2*np.pi,100)
points = np.vstack((np.sin(theta),np.cos(theta)))

fig,ax = plt.subplots(1,figsize=(12,6))
plth,  = ax.plot(np.cos(x),np.sin(x),'ko')
```

The output of `ax.plot` is a variable `plth`, which is a *handle*, or a pointer, to the plot object. That handle allows us to update the locations of the dots rather than redrawing the figure from scratch on each frame.

The second part is to define the function that updates the axis on each frame:

---

2 Swap the minus signs in the sine functions.

```
def aframe(ph):

    # create and apply the transform matrix
    T = np.array([ [1,1-ph],[0,1] ])
    P = T@points

    # update the dots' location
    plth.set_xdata(P[0,:])
    plth.set_ydata(P[1,:])

    return plth
```

Finally, we define our transform parameter ϕ and call the `matplotlib` function that creates the animation:

```
phi = np.linspace(-1,1-1/40,40)**2
animation.FuncAnimation(fig, aframe, phi,
                interval=100, repeat=True)
```

Figure 7-3 show one frame of the movie, and you can watch the entire video by running the code. Admittedly, this movie is unlikely to win any awards, but it does show how matrix multiplication is used in animations. The graphics in CGI movies and video games are slightly more complicated because they use mathematical objects called quaternions, which are vectors in $\mathbb{R}^4$ that allow for rotations and translations in 3D. But the principle—multiply a matrix of geometric coordinates by a transformation matrix—is exactly the same.

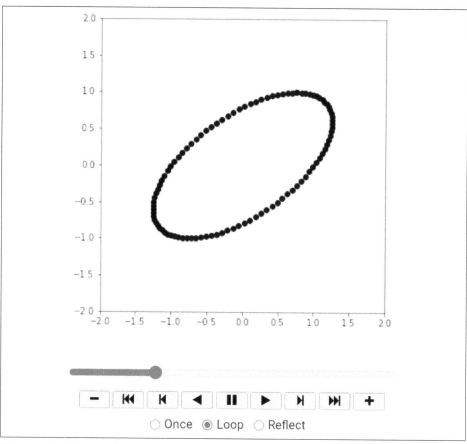

Figure 7-3. A frame of the movie The Wobbly Circle

Before working through the exercises for this section, I encourage you to spend some time playing around with the code for this section. In particular, change the transformation matrix by setting one of the diagonal elements to .5 or 2, change the lower-left off-diagonal element instead of (or in addition to) the upper-right off-diagonal element, parameterize one of the diagonal elements instead of the off-diagonal element, and so on. And here's a question: can you figure out how to get the circle to wobble to the left instead of to the right? The answer is in the footnote.[3]

Image Feature Detection

In this section, I will introduce you to image filtering, which is a mechanism for image feature detection. Image filtering is actually an extension of time series filter-

---

3 Set the lower-right element to −1.

ing, so having gone through Chapter 4 will benefit you here. Recall that to filter or detect features in a time series signal, you design a kernel, and then create a time series of dot products between the kernel and overlapping segments of the signal.

Image filtering works the same way except in 2D instead of 1D. We design a 2D kernel, and then create a new image comprising "dot products" between the kernel and overlapping windows of the image.

I wrote "dot products" in apology quotes because the operation here is not formally the same as the vector dot product. The computation is the same—element-wise multiplication and sum—however, the operation takes place between two matrices, so the implementation is Hadamard multiplication and sum over all matrix elements. Graph A in Figure 7-4 illustrates the procedure. There are additional details of convolution—for example, padding the image so that the result is the same size—that you would learn about in an image-processing book. Here I'd like you to focus on the linear algebra aspects, in particular, the idea that the dot product quantifies the relationship between two vectors (or matrices), which can be used for filtering and feature detection.

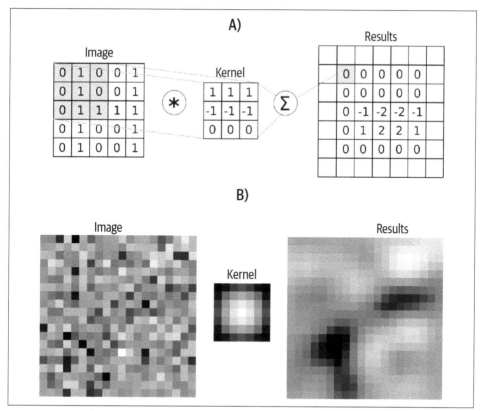

Figure 7-4. Mechanism of image convolution

Before moving on to the analysis, I will briefly explain how a 2D Gaussian kernel is created. A 2D Gaussian is given by the following equation:

$$G = \exp\left(-\left(X^2 + Y^2\right)/\sigma\right)$$

A few notes about that equation: *exp* stands for the natural exponential (the constant $e = 2.71828...$), and $\exp(x)$ is used instead of e^x when the exponential term is long. The X and Y are 2D grids of x,y coordinates on which to evaluate the function. Finally, σ is a parameter of the function that is often called the "shape" or "width": smaller values make the Gaussian narrower, while larger values make the Gaussian wider. For now, I will fix that parameter to certain values, and you'll get to explore the implications of changing that parameter in Exercise 7-6.

Here's how that formula translates into code:

```
Y,X   = np.meshgrid(np.linspace(-3,3,21),np.linspace(-3,3,21))
kernel = np.exp( -(X**2+Y**2) / 20 )
kernel = kernel / np.sum(kernel) # normalize
```

The X and Y grids go from -3 to $+3$ in 21 steps. The width parameter is hard coded to 20. The third line normalizes the values in the kernel so that the sum over the entire kernel is 1. That preserves the original scale of the data. When properly normalized, each step of convolution—and therefore, each pixel in the filtered image —becomes a weighted average of the surrounding pixels, with the weights defined by the Gaussian.

Back to the task at hand: we will smooth a random-numbers matrix, similar to how we smoothed a random-numbers time series in Chapter 4. You can see the random-numbers matrix, the Gaussian kernel, and the result of convolution in Figure 7-4.

The following Python code shows how image convolution is implemented. Again, think back to the time series convolution in Chapter 4 to appreciate that the idea is the same but with an extra dimension necessitating an extra for loop:

```
for rowi in range(halfKr,imgN-halfKr):   # loop over rows
  for coli in range(halfKr,imgN-halfKr): # loop over cols

    # cut out a piece of the image
    pieceOfImg = imagePad[ rowi-halfKr:rowi+halfKr+1:1,
                    coli-halfKr:coli+halfKr+1:1 ]

    # dot product: Hadamard multiply and sum
    dotprod = np.sum( pieceOfImg*kernel )

    # store the result for this pixel
    convoutput[rowi,coli] = dotprod
```

Implementing convolution as a double for loop is actually computationally ineffi-cient. It turns out that convolution can be implemented faster and with less code in the frequency domain. This is thanks to the convolution theorem, which states that convolution in the time (or space) domain is equal to multiplication in the frequency domain. A full exposition of the convolution theorem is outside the scope of this book; I mention it here to justify the recommendation that you use SciPy's convolve2d function instead of the double for loop implementation, particularly for large images.

Let's try smoothing a real picture. We'll use a picture of the Stedelijk Museum in Amsterdam, which I lovingly call "the bathtub from outer space." This image is a 3D matrix because it has rows, columns, and depth—the depth contains pixel intensity values from the red, green, and blue color channels. This picture is stored as a matrix in $\mathbb{R}^{1675 \times 3000 \times 3}$. Formally, that's called a *tensor* because it is a cube, not a spreadsheet, of numbers.

For now, we will reduce it to a 2D matrix by converting to grayscale. That simplifies the computations, although it is not necessary. In Exercise 7-5, you will figure out how to smooth the 3D image. Figure 7-5 shows the grayscale image before and after applying the Gaussian smoothing kernel.

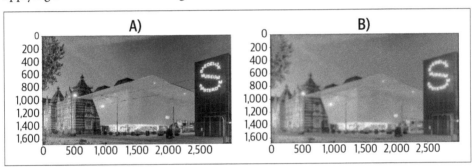

Figure 7-5. A picture of the bathtub museum, before and after a respectable smoothing

Both of these examples used a Gaussian kernel. How many other kernels are avail-able? An infinite number! In Exercise 7-7, you will test two other kernels that are used to identify vertical and horizontal lines. Those feature detectors are common in image processing (and are used by the neurons in your brain to detect edges in the patterns of light that hit your retina).

Image convolution kernels are a major topic in computer vision. In fact, the incred-ible performance of convolutional neural networks (the deep learning architecture optimized for computer vision) is entirely due to the network's ability to craft optimal filter kernels through learning.

Summary

I'll keep this simple: the point is that—yet again—incredibly important and sophisticated methods in data science and machine learning are built on simple linear algebra principles.

Code Exercises

Covariance and Correlation Matrices Exercises

Exercise 7-1.

In this exercise, you will transform the covariance matrix into a correlation matrix. The procedure involves dividing each matrix element (that is, the covariance between each pair of variables) by the product of the variances of those two variables.

This is implemented by pre- and postmultiplying the covariance matrix by a diagonal matrix containing inverted standard deviations of each variable (standard deviation is the square root of variance). The standard deviations are inverted because we need to *divide* by the variances although we will *multiply* matrices. The reason for pre- and postmultiplying by standard deviations is the special property of pre- and postmultiplying by a diagonal matrix, which was explained in Exercise 5-11.

Equation 7-1 shows the formula.

Exercise 7-1. Correlation from covariance

$$\mathbf{R} = \mathbf{SCS}$$

C is the covariance matrix, and **S** is a diagonal matrix of reciprocated standard deviations per variable (that is, the ith diagonal is $1/\sigma_i$ where σ_i is the standard deviation of variable i).

Your goal in this exercise is to compute the correlation matrix from the covariance matrix, by translating Equation 7-1 into Python code. You can then reproduce Figure 7-6.

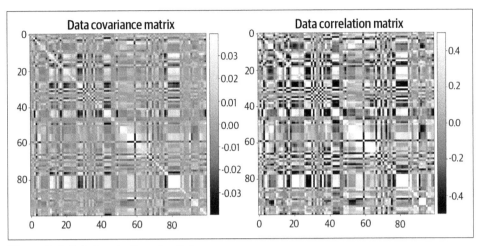

Figure 7-6. Solution to Exercise 7-1

Exercise 7-2.

NumPy has a function `np.corrcoef()` that returns a correlation matrix, given an input data matrix. Use this function to reproduce the correlation matrix you created in the previous exercise. Show both matrices, and their difference, in a figure like Figure 7-7 to confirm that they are the same.

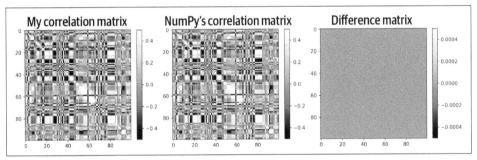

Figure 7-7. Solution to Exercise 7-2. Note the difference in color scaling.

Next, inspect the source code of `np.corrcoef()` by evaluating `??np.corrcoef()`. NumPy uses a slightly different implementation of broadcast dividing by the standard deviations instead of pre- and postmultiplying by a diagonal matrix of inverted standard deviations, but you should be able to understand how their code implementation matches the math and the Python code you wrote in the previous exercise.

Geometric Transformations Exercises

Exercise 7-3.

The goal of this exercise is to show points in a circle before and after applying a transformation, similar to how I showed the line before and after rotation in Figure 7-2. Use the following transformation matrix and then create a graph that looks like Figure 7-8:

$$\mathbf{T} = \begin{bmatrix} 1 & .5 \\ 0 & .5 \end{bmatrix}$$

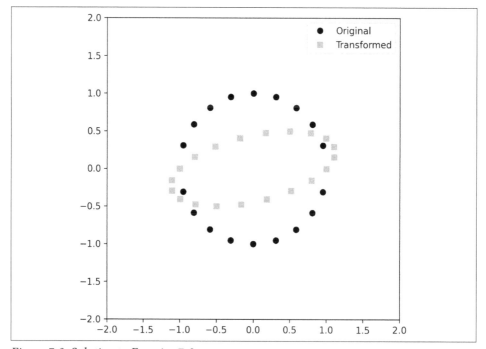

Figure 7-8. Solution to Exercise 7-3

Exercise 7-4.

Now for another movie. I call this one *The Coiling DNA*. Figure 7-9 shows one frame of the movie. The procedure is the same as for *The Wobbly Circle*—set up a figure, create a Python function that applies a transformation matrix to a matrix of coordinates, and tell matplotlib to create an animation using that function. Use the following transformation matrix:

$$T = \begin{bmatrix} (1 - \phi/3) & 0 \\ 0 & \phi \end{bmatrix}$$

$-1 \leq \phi \leq 1$

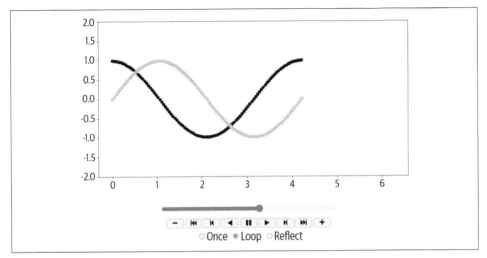

Figure 7-9. Solution to Exercise 7-4

Image Feature Detection Exercises

Exercise 7-5.

Smooth the 3D bathtub picture (if you need a hint, check the footnote[4]).

The output of the convolve2d function has a data type float64 (you can see this yourself by typing variableName.dtype). However, plt.imshow will give a warning about clipping numerical values, and the picture won't render properly. Therefore, you'll need to convert the result of convolution to uint8.

Exercise 7-6.

You don't need to use the same kernel for each color channel. Change the width parameter of the Gaussian for each channel according to the values shown in Figure 7-10. The effect on the image is subtle, but the different blurs of the different colors give it a bit of a 3D look, as if you are looking at a red-blue anaglyph without the glasses.

---

4 Hint: smooth each color channel separately.

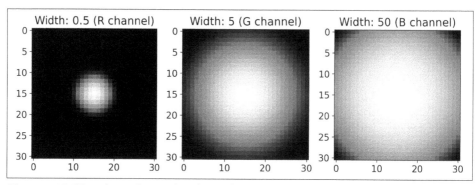

Figure 7-10. Kernels used per color channel in Exercise 7-6

Exercise 7-7.

Technically, image smoothing is feature extraction, because it involves extracting the smooth features of the signal while dampening the sharp features. Here we will change the filter kernels to solve other image feature detection problems: identifying horizontal and vertical lines.

The two kernels are shown in Figure 7-11, as are their effects on the image. You can handcraft the two kernels based on their visual appearance; they are 3×3 and comprise only the numbers -1, 0, and $+1$. Convolve those kernels with the 2D grayscale picture to create the feature maps shown in Figure 7-11.

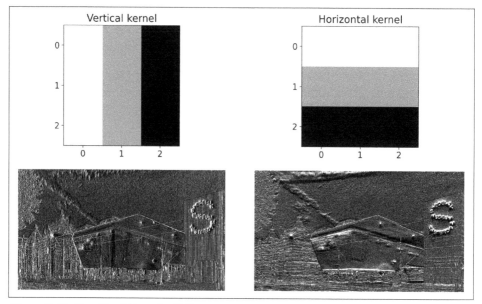

Figure 7-11. Results of Exercise 7-7

## CHAPTER 8

Matrix Inverse

We are moving toward solving matrix equations. Matrix equations are like regular equations (e.g., solve for x in $4x = 8$) but…they have matrices. By this point in the book, you are well aware that things get complicated when matrices get involved. Nonetheless, we must embrace that complexity, because solving matrix equations is a huge part of data science.

The matrix inverse is central to solving matrix equations in practical applications, including fitting statistical models to data (think of general linear models and regression). By the end of this chapter, you will understand what the matrix inverse is, when it can and cannot be computed, how to compute it, and how to interpret it.

The Matrix Inverse

The inverse of matrix $\mathbf{A}$ is another matrix $\mathbf{A}^{-1}$ (pronounced "A inverse") that multiplies $\mathbf{A}$ to produce the identity matrix. In other words, $\mathbf{A}^{-1}\mathbf{A} = \mathbf{I}$. That is how you "cancel" a matrix. Another conceptualization is that we want to linearly transform a matrix into the identity matrix; the matrix inverse contains that linear transformation, and matrix multiplication is the mechanism of applying that transformation to the matrix.

But why do we even need to invert matrices? We need to "cancel" a matrix in order to solve problems that can be expressed in the form $\mathbf{A}\mathbf{x} = \mathbf{b}$, where $\mathbf{A}$ and $\mathbf{b}$ are known quantities and we want to solve for $\mathbf{x}$. The solution has the following general form:

$$\mathbf{A}\mathbf{x} = \mathbf{b}$$
$$\mathbf{A}^{-1}\mathbf{A}\mathbf{x} = \mathbf{A}^{-1}\mathbf{b}$$
$$\mathbf{I}\mathbf{x} = \mathbf{A}^{-1}\mathbf{b}$$
$$\mathbf{x} = \mathbf{A}^{-1}\mathbf{b}$$

That seems really straightforward, but computing the inverse is deceptively difficult, as you will soon learn.

Types of Inverses and Conditions for Invertibility

"Invert the matrix" sounds like it should always work. Who wouldn't want to invert a matrix whenever it's convenient? Unfortunately, life is not always so simple: not all matrices can be inverted.

There are three different kinds of inverses that have different conditions for invertibility. They are introduced here; details are in subsequent sections:

Full inverse

This means $\mathbf{A}^{-1}\mathbf{A} = \mathbf{A}\mathbf{A}^{-1} = \mathbf{I}$. There are two conditions for a matrix to have a full inverse: (1) square and (2) full-rank. Every square full-rank matrix has an inverse, and every matrix that has a full inverse is square and full-rank. By the way, I'm using the term *full inverse* here to distinguish it from the next two possibilities; you would typically refer to the full inverse simply as the *inverse*.

One-sided inverse

A one-sided inverse can transform a rectangular matrix into the identity matrix, but it works only for one multiplication order. In particular, a tall matrix $\mathbf{T}$ can have a *left-inverse*, meaning $\mathbf{L}\mathbf{T} = \mathbf{I}$ but $\mathbf{T}\mathbf{L} \neq \mathbf{I}$. And a wide matrix $\mathbf{W}$ can have a *right-inverse*, meaning that $\mathbf{W}\mathbf{R} = \mathbf{I}$ but $\mathbf{R}\mathbf{W} \neq \mathbf{I}$.

A nonsquare matrix has a one-sided inverse only if it has the maximum possible rank. That is, a tall matrix has a left-inverse if it is rank-N (full column rank) while a wide matrix has a right-inverse if it is rank-M (full row rank).

Pseudoinverse

Every matrix has a pseudoinverse, regardless of its shape and rank. If the matrix is square full-rank, then its pseudoinverse equals its full inverse. Likewise, if the matrix is nonsquare and has its maximum possible rank, then the pseudoinverse equals its left-inverse (for a tall matrix) or its right-inverse (for a wide matrix). But a reduced-rank matrix still has a pseudoinverse, in which case the pseudoinverse transforms the singular matrix into another matrix that is close but not equal to the identity matrix.

Matrices that do not have a full or one-sided inverse are called *singular* or *noninvertible*. That is the same thing as labeling a matrix *reduced-rank* or *rank-deficient*.

Computing the Inverse

The matrix inverse sounds great! How do we compute it? Let's start by thinking about how to compute the *scalar* inverse: you simply invert (take the reciprocal of) the number. For example, the inverse of the number 3 is 1/3, which is the same thing as 3^{-1}. Then, $3 \times 3^{-1} = 1$.

Based on this reasoning, you might guess that the matrix inverse works the same way: invert each matrix element. Let's try:

$$\begin{bmatrix} a & b \\ c & d \end{bmatrix}^{-1} = \begin{bmatrix} 1/a & 1/b \\ 1/c & 1/d \end{bmatrix}$$

Unfortunately, this does not produce the desired outcome, which is readily demonstrated by multiplying the original matrix by the matrix of individually inverted elements:

$$\begin{bmatrix} a & b \\ c & d \end{bmatrix}\begin{bmatrix} 1/a & 1/b \\ 1/c & 1/d \end{bmatrix} = \begin{bmatrix} 1 + b/c & a/b + b/d \\ c/a + d/c & 1 + c/b \end{bmatrix}$$

That is a valid multiplication, but it does *not* produce the identity matrix, which means that the matrix with individual elements inverted is *not* the matrix inverse.

There is an algorithm for computing the matrix of any invertible matrix. It's long and tedious (this is why we have computers do the number crunching for us!), but there are a few shortcuts for special matrices.

Inverse of a 2 × 2 Matrix

To invert a 2 × 2 matrix, swap the diagonal elements, multiply the off-diagonal elements by −1, and divide by the determinant. That algorithm will produce a matrix that transforms the original matrix into the identity matrix.

Observe:

$$\mathbf{A} = \begin{bmatrix} a & b \\ c & d \end{bmatrix}$$

$$\mathbf{A}^{-1} = \frac{1}{ad - bc}\begin{bmatrix} d & -b \\ -c & a \end{bmatrix}$$

$$\mathbf{AA}^{-1} = \begin{bmatrix} a & b \\ c & d \end{bmatrix}\frac{1}{ad - bc}\begin{bmatrix} d & -b \\ -c & a \end{bmatrix}$$

$$= \frac{1}{ad - bc}\begin{bmatrix} ad - bc & 0 \\ 0 & ad - bc \end{bmatrix}$$

$$= \begin{bmatrix} 1 & 0 \\ 0 & 1 \end{bmatrix}$$

Let's work through a numerical example:

$$\begin{bmatrix} 1 & 4 \\ 2 & 7 \end{bmatrix}\begin{bmatrix} 7 & -4 \\ -2 & 1 \end{bmatrix}\frac{1}{7-8} = \begin{bmatrix} (7-8) & (-4+4) \\ (14-14) & (-8+7) \end{bmatrix}\frac{1}{-1} = \begin{bmatrix} 1 & 0 \\ 0 & 1 \end{bmatrix}$$

That worked out nicely.

Computing the inverse in Python is easy:

```
A = np.array([ [1,4],[2,7] ])
Ainv = np.linalg.inv(A)
A@Ainv
```

You can confirm that A@Ainv gives the identity matrix, as does Ainv@A. Of course, A*Ainv does not give the identity matrix, because * is for Hadamard (element-wise) multiplication.

The matrix, its inverse, and their product are visualized in Figure 8-1.

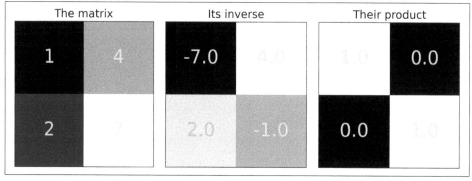

Figure 8-1. Inverse of a 2 × 2 matrix

Let's try another example:

$$\begin{bmatrix} 1 & 4 \\ 2 & 8 \end{bmatrix} \begin{bmatrix} 8 & -4 \\ -2 & 1 \end{bmatrix} \frac{1}{0} = \begin{bmatrix} (8-8) & (-4+4) \\ (16-16) & (-8+8) \end{bmatrix} \frac{1}{0} = \text{ ???}$$

There are several problems with this example. The matrix multiplication gives us **0** instead of $\Delta\mathbf{I}$. But there is a bigger problem—the determinant is zero! Mathematicians have been warning us for ages that we cannot divide by zero. Let's not start doing it now.

What's different about the second example? It is a reduced-rank matrix (rank = 1). This shows a numerical example that reduced-rank matrices are not invertible.

What does Python do in this case? Let's find out:

```
A = np.array([ [1,4],[2,8] ])
Ainv = np.linalg.inv(A)
A@Ainv
```

Python won't even try to calculate the result like I did. Instead, it gives an error with the following message:

```
LinAlgError: Singular matrix
```

Reduced-rank matrices do not have an inverse, and programs like Python won't even try to calculate one. However, this matrix does have a pseudoinverse. I'll get back to that in a few sections.

Inverse of a Diagonal Matrix

There is also a shortcut to compute the inverse of a square diagonal matrix. The insight that leads to this shortcut is that the product of two diagonal matrices is simply the diagonal elements scalar multiplied (discovered in Exercise 5-12). Consider the example below; before continuing with the text, try to figure out the shortcut for the inverse of a diagonal matrix:

$$\begin{bmatrix} 2 & 0 & 0 \\ 0 & 3 & 0 \\ 0 & 0 & 4 \end{bmatrix} \begin{bmatrix} b & 0 & 0 \\ 0 & c & 0 \\ 0 & 0 & d \end{bmatrix} = \begin{bmatrix} 2b & 0 & 0 \\ 0 & 3c & 0 \\ 0 & 0 & 4d \end{bmatrix}$$

Have you figured out the trick for computing the inverse of a diagonal matrix? The trick is that you simply invert each diagonal element, while ignoring the off-diagonal zeros. That's clear in the previous example by setting $b = 1/2$, $c = 1/3$, and $d = 1/4$.

What happens when you have a diagonal matrix with a zero on the diagonal? You cannot invert that element because you'll have 1/0. So a diagonal matrix with at least one zero on the diagonal has no inverse. (Also remember from Chapter 6 that a diagonal matrix is full-rank only if all diagonal elements are nonzero.)

The inverse of a diagonal matrix is important because it directly leads to the formula for computing the pseudoinverse. More on this later.

Inverting Any Square Full-Rank Matrix

To be honest, I debated whether to include this section. The full algorithm for inverting an invertible matrix is long and tedious, and you will never need to use it in applications (instead, you will use np.linalg.inv or other functions that call inv).

On the other hand, implementing the algorithm in Python is an excellent opportunity for you to practice your skills of translating an algorithm, described in equations and English, into Python code. Therefore, I will explain how the algorithm works here without showing any code. I encourage you to code the algorithm in Python as you read this section, and you can check your solution against mine in Exercise 8-2 in the online code.

The algorithm to compute the inverse involves four intermediate matrices, called the *minors, grid, cofactors,* and *adjugate matrices*:

The minors matrix

This matrix comprises determinants of submatrices. Each element $m_{i,j}$ of the minors matrix is the determinant of the submatrix created by excluding the *i*th row and the *j*th column. Figure 8-2 shows an overview of the procedure for a 3×3 matrix.

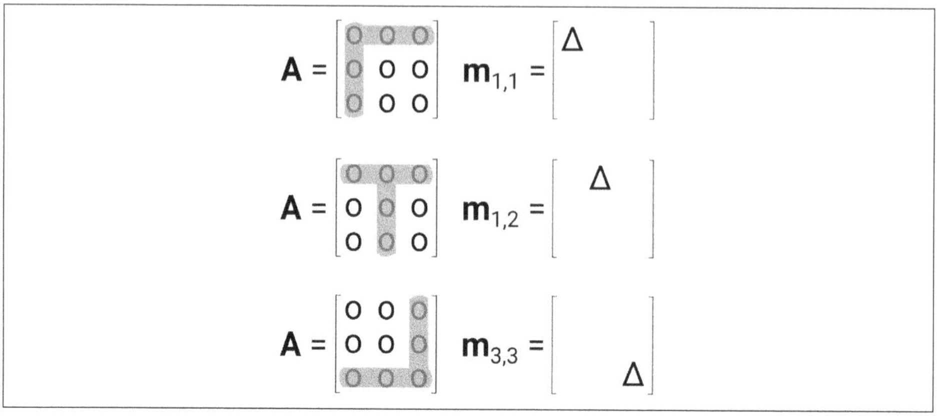

Figure 8-2. Computing the minors matrix (gray shaded areas are removed to create each submatrix)

The grid matrix

The grid matrix is a checkerboard of alternating +1s and −1s. It is computed using the following formula:

$$g_{i,j} = -1^{i+j}$$

Be careful with the indexing and exponentiating when implementing that formula in Python. You should inspect the matrix carefully to make sure it is a checkerboard of alternating signs, with +1 in the top-left element.

The cofactors matrix

The cofactors matrix is the Hadamard multiplication of the minors matrix with the grid matrix.

Adjugate matrix

This is the transpose of the cofactors matrix, scalar multiplied by the inverse of the determinant of the original matrix (the matrix you are computing the inverse of, not the cofactors matrix).

The adjugate matrix is the inverse of the original matrix.

Figure 8-3 shows the four intermediate matrices, the inverse returned by np.linalg.inv, and the identity matrix resulting from the multiplication of the original matrix with its inverse computed according to the procedure previously described. The original matrix was a random-numbers matrix.

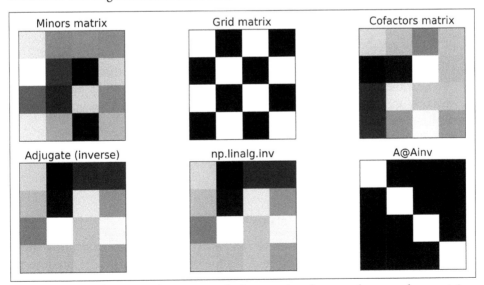

Figure 8-3. Visualizations of inverse-producing matrices for a random-numbers matrix

One-Sided Inverses

A tall matrix does not have a full inverse. That is, for matrix $\mathbf{T}$ of size $M > N$, there is no tall matrix $\mathbf{T}^{-1}$ such that $\mathbf{TT}^{-1} = \mathbf{T}^{-1}\mathbf{T} = \mathbf{I}$.

But there is a matrix $\mathbf{L}$ such that $\mathbf{LT} = \mathbf{I}$. Our goal now is to find that matrix. We start by making matrix $\mathbf{T}$ square. How do we make a nonsquare matrix into a square matrix? Of course you know the answer—we multiply it by its transpose.

Here's the next question: should we compute $\mathbf{T}^T\mathbf{T}$ or $\mathbf{TT}^T$? Both are square...but $\mathbf{T}^T\mathbf{T}$ is full-rank if $\mathbf{T}$ is full column-rank. And why is that important? You guessed it—all square full-rank matrices have an inverse. Before presenting the derivation of the left-inverse, let's demonstrate in Python that a tall matrix times its transpose has a full inverse:

```
T = np.random.randint(-10,11,size=(40,4))
TtT = T.T @ T
TtT_inv = np.linalg.inv(TtT)
TtT_inv @ TtT
```

You can confirm in code that the final line produces the identity matrix (within machine-precision error).

Let me translate that Python code into a mathematical equation:

$$\left(\mathbf{T}^T\mathbf{T}\right)^{-1}\left(\mathbf{T}^T\mathbf{T}\right) = \mathbf{I}$$

From the code and formula, you can see that because $\mathbf{T}^T\mathbf{T}$ is not the same matrix as $\mathbf{T}$, $\left(\mathbf{T}^T\mathbf{T}\right)^{-1}$ is *not* the inverse of $\mathbf{T}$.

But—and here is the key insight—we're looking for a matrix that left-multiplies $\mathbf{T}$ to produce the identity matrix; we don't actually care what other matrices need to multiply to produce that matrix. So let's break apart and regroup the matrix multiplications:

$$\mathbf{L} = \left(\mathbf{T}^T\mathbf{T}\right)^{-1}\mathbf{T}^T$$
$$\mathbf{LT} = \mathbf{I}$$

That matrix $\mathbf{L}$ is the *left-inverse* of matrix $\mathbf{T}$.

Now we can finish the Python code to compute the left-inverse and confirm that it conforms to our specification. Left-multiply the original tall matrix to produce the identity matrix:

```
L = TtT_inv @ T.T # the left-inverse
L@T # produces the identity matrix
```

You can also confirm in Python that **TL** (that is, *right*-multiplying by the left-inverse) is *not* the identity matrix. That's why the left-inverse is a one-sided inverse.

Figure 8-4 illustrates a tall matrix, its left-inverse, and the two ways of multiplying the left-inverse by the matrix. Note that the left-inverse is not a square matrix and that postmultiplying by the left-inverse gives a result that is definitely not the identity matrix.

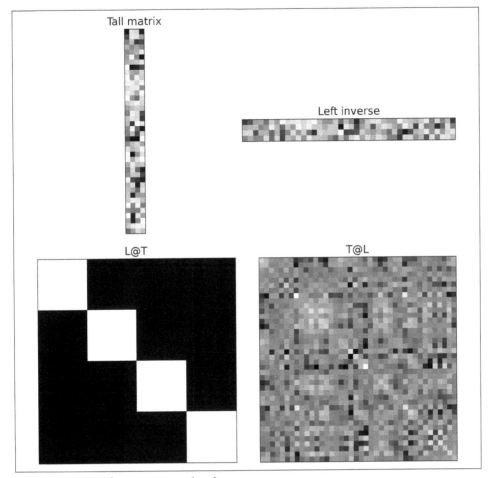

Figure 8-4. The left-inverse, visualized

The left-inverse is extremely important. In fact, once you learn about fitting statistical models to data and the least squares solution, you'll see the left-inverse all over the place. It is no hyperbole to state that the left-inverse is one of the most important contributions of linear algebra to modern human civilization.

One final note about the left-inverse, which was implicit in this discussion: the left-inverse is defined only for tall matrices that have full column-rank. A matrix of size $M > N$ with rank $r < N$ does not have a left-inverse. Why is that? The answer is in the footnote.[1]

Now you know how to compute the left-inverse. What about the right-inverse? I refuse to teach you how to compute it! That's not because it's secret knowledge that I'm keeping from you, and it's certainly not because I don't like you. Instead, it's because I want you to challenge yourself to derive the math of the right-inverse and demonstrate it in code using Python. Exercise 8-4 has a few hints if you need them, or you can try to figure it out now before continuing to the next section.

The Inverse Is Unique

The matrix inverse is unique, meaning that if a matrix has an inverse, it has exactly one inverse. There cannot be two matrices $\mathbf{B}$ and $\mathbf{C}$ such that $\mathbf{AB} = \mathbf{I}$ and $\mathbf{AC} = \mathbf{I}$ while $\mathbf{B} \neq \mathbf{C}$.

There are several proofs of this claim. The one I'll show relies on a technique called *proof by negation*. That means that we try but fail to prove a false claim, thereby proving the correct claim. In this case, we start with three assumptions: (1) matrix $\mathbf{A}$ is invertible, (2) matrices $\mathbf{B}$ and $\mathbf{C}$ are inverses of $\mathbf{A}$, and (3) matrices $\mathbf{B}$ and $\mathbf{C}$ are distinct, meaning $\mathbf{B} \neq \mathbf{C}$. Follow each expression from left to right, and notice that each subsequent expression is based on adding or removing the identity matrix, expressed as the matrix times its inverse:

$$\mathbf{C} = \mathbf{CI} = \mathbf{CAB} = \mathbf{IB} = \mathbf{B}$$

All statements are equal, which means the first and final expressions are equal, which means that our assumption of $\mathbf{B} \neq \mathbf{C}$ is false. The conclusion is that any two matrices that claim to be the inverse of the same matrix are equal. In other words, an invertible matrix has exactly one inverse.

Moore-Penrose Pseudoinverse

As I wrote earlier, it is simply impossible to transform a reduced-rank matrix into the identity matrix via matrix multiplication. That means that reduced-rank matrices do not have a full or a one-sided inverse. But singular matrices do have pseudoinverses. Pseudoinverses are transformation matrices that bring a matrix close to the identity matrix.

---

1 Because $\mathbf{T}^T\mathbf{T}$ is reduced-rank and thus cannot be inverted.

The plural *pseudoinverses* was not a typo: although the full matrix inverse is unique, the pseudoinverse is not unique. A reduced-rank matrix has an infinite number of pseudoinverses. But some are better than others, and there is really only one pseudoinverse worth discussing, because it is most likely the only one you'll ever use.

That is called the *Moore-Penrose pseudoinverse*, sometimes abbreviated as the MP pseudoinverse. But because this is by far the most commonly used pseudoinverse, you can always assume that the term *pseudoinverse* refers to the Moore-Penrose pseudoinverse.

The following matrix is the pseudoinverse of the singular matrix you saw earlier in this chapter. The first line shows the pseudoinverse of the matrix, and the second line shows the product of the matrix and its pseudoinverse:

$$\begin{bmatrix} 1 & 4 \\ 2 & 8 \end{bmatrix}^{\dagger} = \frac{1}{85}\begin{bmatrix} 1 & 2 \\ 4 & 8 \end{bmatrix}$$

$$\begin{bmatrix} 1 & 4 \\ 2 & 8 \end{bmatrix} \frac{1}{85}\begin{bmatrix} 1 & 2 \\ 4 & 8 \end{bmatrix} = \begin{bmatrix} .2 & .4 \\ .4 & .8 \end{bmatrix}$$

(The scaling factor of 85 was extracted to facilitate visual inspection of the matrix.)

The pseudoinverse is indicated using a dagger, a plus sign, or an asterisk in the superscript: $\mathbf{A}^{\dagger}$, $\mathbf{A}^{+}$, or $\mathbf{A}^{*}$.

The pseudoinverse is implemented in Python using the function np.linalg.pinv. The following code computes the pseudoinverse of the singular matrix for which np.linalg.inv produced an error message:

```
A = np.array([ [1,4],[2,8] ])
Apinv = np.linalg.pinv(A)
A@Apinv
```

How is the pseudoinverse computed? The algorithm is either incomprehensible or intuitive, depending on whether you understand the singular value decomposition. I will explain the pseudoinverse computation briefly, but if you don't understand it, then please don't worry: I promise it will be intuitive by the end of Chapter 13. To compute the pseudoinverse, take the SVD of a matrix, invert the nonzero singular values without changing the singular vectors, and reconstruct the matrix by multiplying $\mathbf{U}\mathbf{\Sigma}^{+}\mathbf{V}^{\mathrm{T}}$.

Numerical Stability of the Inverse

Computing the matrix inverse involves a lot of FLOPs (floating-point operations), including many determinants. You learned in Chapter 6 that the determinant of a matrix can be numerically unstable, and therefore computing *many* determinants can lead to numerical inaccuracies, which can accumulate and cause significant problems when working with large matrices.

For this reason, the low-level libraries that implement numerical computations (for example, LAPACK) generally strive to avoid explicitly inverting matrices when possible, or they decompose matrices into the product of other matrices that are more numerically stable (e.g., QR decomposition, which you will learn about in Chapter 9).

Matrices that have numerical values in roughly the same range tend to be more stable (though this is not guaranteed), which is why random-numbers matrices are easy to work with. But matrices with a large range of numerical values have a high risk of numerical instability. The "range of numerical values" is more formally captured as the *condition number* of a matrix, which is the ratio of the largest to smallest singular value. You'll learn more about condition number in Chapter 14; for now, suffice it to say that the condition number is a measure of the spread of numerical values in a matrix.

An example of a numerically unstable matrix is a Hilbert matrix. Each element in a Hilbert matrix is defined by the simple formula shown in Equation 8-1.

Equation 8-1. Formula to create a Hilbert matrix. i and j are row and column indices.

$$h_{i,j} = \frac{1}{i+j-1}$$

Here's an example of a 3 × 3 Hilbert matrix:

$$\begin{bmatrix} 1 & 1/2 & 1/3 \\ 1/2 & 1/3 & 1/4 \\ 1/3 & 1/4 & 1/5 \end{bmatrix}$$

As the matrix gets larger, the range of numerical values increases. As a consequence, the computer-calculated Hilbert matrix quickly becomes rank-deficient. Even full-rank Hilbert matrices have inverses in a very different numerical range. This is illustrated in Figure 8-5, which shows a 5 × 5 Hilbert matrix, its inverse, and their product.

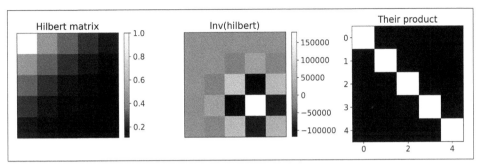

Figure 8-5. A Hilbert matrix, its inverse, and their product

The product matrix certainly looks like the identity matrix, but in Exercise 8-9 you will discover that looks can be deceiving, and rounding errors increase dramatically with matrix size.

Geometric Interpretation of the Inverse

In Chapters 6 and 7, you learned how to conceptualize matrix-vector multiplication as a geometric transformation of a vector or a set of points.

Following along these lines, we can think of the matrix inverse as *undoing* the geometric transformation imposed by matrix multiplication. Figure 8-6 shows an example that follows from Figure 7-8; I simply multiplied the transformed geometric coordinates by the inverse of the transformation matrix.

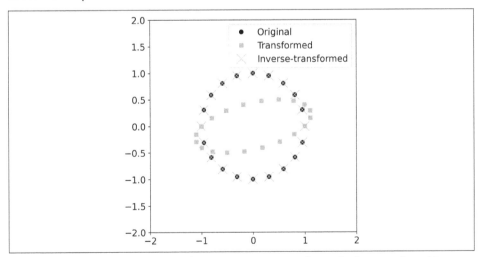

Figure 8-6. The matrix inverse undoes a geometric transform. Code to produce this figure is part of Exercise 8-8.

This geometric effect is unsurprising when inspecting the math. In the following equations, $\mathbf{P}$ is the $2 \times N$ matrix of original geometric coordinates, $\mathbf{T}$ is the transformation matrix, $\mathbf{Q}$ is the matrix of transformed coordinates, and $\mathbf{U}$ is the matrix of back-transformed coordinates:

$$\mathbf{Q} = \mathbf{TP}$$
$$\mathbf{U} = \mathbf{T}^{-1}\mathbf{Q}$$
$$\mathbf{U} = \mathbf{T}^{-1}\mathbf{TP}$$

Although not a surprising result, I hope it helps build some geometric intuition for the purpose of the matrix inverse: *undoing the transform imposed by the matrix*. This interpretation will come in handy when you learn about diagonalizing a matrix through eigendecomposition.

This geometric interpretation also provides some intuition for why a reduced-rank matrix has no inverse: the geometric effect of transforming by a singular matrix is that at least one dimension is flattened. Once a dimension is flattened, it cannot be unflattened, just like you cannot see your back when facing a mirror.[2]

Summary

I really enjoyed writing this chapter, and I hope you enjoyed learning from it. Here is a summary of the key take-home messages:

- The matrix inverse is a matrix that, through matrix multiplication, transforms a maximum-rank matrix into the identity matrix. The inverse has many purposes, including moving matrices around in an equation (e.g., solve for $\mathbf{x}$ in $\mathbf{Ax} = \mathbf{b}$).

- A square full-rank matrix has a full inverse, a tall full column-rank matrix has a left-inverse, and a wide full row-rank matrix has a right-inverse. Reduced-rank matrices cannot be linearly transformed into the identity matrix, but they do have a pseudoinverse that transforms the matrix into another matrix that is closer to the identity matrix.

- The inverse is unique—if a matrix can be linearly transformed into the identity matrix, there is only one way to do it.

- There are some tricks for computing the inverses of some kinds of matrices, including 2×2 and diagonal matrices. These shortcuts are simplifications of the full formula for computing a matrix inverse.

---

2 There's a witty analogy to *Flatland* somewhere here that I'm not quite able to eloquate. The point is: read the book *Flatland*.

- Due to the risk of numerical precision errors, production-level algorithms try to avoid explicitly inverting matrices, or will decompose a matrix into other matrices that can be inverted with greater numerical stability.

Code Exercises

Exercise 8-1.

The inverse of the inverse is the original matrix; in other words, $\left(\mathbf{A}^{-1}\right)^{-1} = \mathbf{A}$. This is analgous to how $1/(1/a) = a$. Illustrate this using Python.

Exercise 8-2.

Implement the full algorithm described in "Inverting Any Square Full-Rank Matrix" on page 134 and reproduce Figure 8-3. Of course, your matrices will look different from Figure 8-3 because of random numbers, although the grid and identity matrices will be the same.

Exercise 8-3.

Implement the full-inverse algorithm by hand for a 2×2 matrix using matrix elements a, b, c, and d. I don't normally assign hand-solved problems in this book, but this exercise will show you where the shortcut comes from. Remember that the determinant of a scalar is its absolute value.

Exercise 8-4.

Derive the right-inverse for wide matrices by following the logic that allowed us to discover the left-inverse. Then reproduce Figure 8-4 for a wide matrix. (Hint: start from the code for the left-inverse and adjust as necessary.)

Exercise 8-5.

Illustrate in Python that the pseudoinverse (via `np.linalg.pinv`) equals the full inverse (via `np.linalg.inv`) for an invertible matrix. Next, illustrate that the pseudoinverse equals the left-inverse for a tall full column-rank matrix, and that it equals the right-inverse for a wide full row-rank matrix.

Exercise 8-6.

The LIVE EVIL rule applies to the inverse of multiplied matrices. Test this in code by creating two square full-rank matrices $\mathbf{A}$ and $\mathbf{B}$, then use Euclidean distance to compare (1) $(\mathbf{AB})^{-1}$, (2) $\mathbf{A}^{-1}\mathbf{B}^{-1}$, and (3) $\mathbf{B}^{-1}\mathbf{A}^{-1}$. Before starting to code, make a prediction about which results will be equal. Print out your results using formatting like the following (I've omitted my results so you won't be biased!):

```
Distance between (AB)^-1 and (A^-1)(B^-1) is ___
Distance between (AB)^-1 and (B^-1)(A^-1) is ___
```

As an extra challenge, you can confirm that the LIVE EVIL rule applies to a longer string of matrices, e.g., four matrices instead of two.

Exercise 8-7.

Does the LIVE EVIL rule also apply to the one-sided inverse? That is, does $\left(\mathbf{T}^{\mathsf{T}}\mathbf{T}\right)^{-1} = \mathbf{T}^{-\mathsf{T}}\mathbf{T}^{-1}$? As with the previous exercise, make a prediction and then test it in Python.

Exercise 8-8.

Write code to reproduce Figure 8-6. Start by copying the code from Exercise 7-3. After reproducing the figure, make the transformation matrix noninvertible by setting the lower-left element to 1. What else needs to be changed in the code to prevent errors?

Exercise 8-9.

This and the next exercise will help you explore the matrix inverse and its risk of numerical instability, using the Hilbert matrix. Start by creating a Hilbert matrix. Write a Python function that takes an integer as input and produces a Hilbert matrix as output, following Equation 8-1. Then reproduce Figure 8-5.

I recommend writing your Python function using a double for loop over the rows and columns (*i* and *j* matrix indices), following the math formula. Once you confirm that the function is accurate, you can optionally challenge yourself to rewrite the function without any for loops (hint: outer product). You can confirm the accuracy of your function by comparing it to the hilbert function, which is in the scipy.linalg library.

Exercise 8-10.

Using your Hilbert matrix function, create a Hilbert matrix, then compute its inverse using np.linalg.inv, and compute the product of the two matrices. That product should equal the identity matrix, which means the Euclidean distance between that product and the true identity matrix produced by `np.eye` should be 0 (within computer rounding error). Compute that Euclidean distance.

Put this code into a `for` loop over a range of matrix sizes ranging from 3×3 to 12×12. For each matrix size, store the Euclidean distance and the condition number of the Hilbert matrix. As I wrote earlier, the condition number is a measure of the spread of numerical values in the matrix, and can be extracted using the function `np.linalg.cond`.

Next, repeat the previous code, but using a Gaussian random-numbers matrix instead of the Hilbert matrix.

Finally, plot all of the results as shown in Figure 8-7. I plotted the distance and condition number in log scale to facilitate visual interpretation.

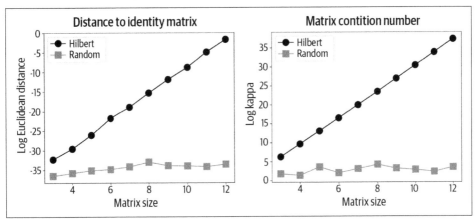

Figure 8-7. Results of Exercise 8-10

Please feel inspired to continue exploring linear algebra using this exercise! Try plotting the Hilbert matrix times its inverse (consider adjusting the color scaling), using larger matrices or different special matrices, extracting other properties of the matrix like rank or norm, etc. You're on an adventure in the wonderful land of linear algebra, and Python is the magic carpet that whisks you across the landscape.

## CHAPTER 9

Orthogonal Matrices and QR Decomposition

You will learn five major decompositions in this book: orthogonal vector decomposition, QR decomposition, LU decomposition, eigendecomposition, and singular value decomposition. Those are not the only decompositions in linear algebra, but they are the most important ones for data science and machine learning.

In this chapter, you will learn QR. And along the way, you'll learn a new special type of matrix (orthogonal). QR decomposition is a workhorse that powers applications including the matrix inverse, least squares model fitting, and eigendecomposition. Therefore, understanding and gaining familiarity with QR decomposition will help you level up your linear algebra skills.

Orthogonal Matrices

I will begin by introducing you to orthogonal matrices. An *orthogonal matrix* is a special matrix that is important for several decompositions, including QR, eigendecomposition, and singular value decomposition. The letter **Q** is often used to indicate orthogonal matrices. Orthogonal matrices have two properties:

Orthogonal columns
    All columns are pair-wise orthogonal.

Unit-norm columns
    The norm (geometric length) of each column is exactly 1.

We can translate those two properties into a mathematical expression (remember that $\langle \mathbf{a}, \mathbf{b} \rangle$ is an alternative notation for the dot product):

$$\langle \mathbf{q}_i, \mathbf{q}_j \rangle = \begin{cases} 0, & \text{if } i \neq j \\ 1, & \text{if } i = j \end{cases}$$

What does that mean? It means that the dot product of a column with itself is 1 while the dot product of a column with any other column is 0. That's a lot of dot products with only two possible outcomes. We can organize all of the dot products amongst all pairs of columns by premultiplying the matrix by its transpose. Remember that matrix multiplication is defined as dot products between all rows of the left matrix with all columns of the right matrix; therefore, the rows of $\mathbf{Q}^T$ are the columns of $\mathbf{Q}$.

The matrix equation expressing the two key properties of an orthogonal matrix is simply marvelous:

$$\mathbf{Q}^T\mathbf{Q} = \mathbf{I}$$

The expression $\mathbf{Q}^T\mathbf{Q} = \mathbf{I}$ is amazing. Really, it's a big deal.

Why is it a big deal? Because $\mathbf{Q}^T$ is a matrix that multiplies $\mathbf{Q}$ to produce the identity matrix. That's the exact same definition as the matrix inverse. Thus, the inverse of an orthogonal matrix is its transpose. That's crazy cool, because the matrix inverse is tedious and prone to numerical inaccuracies, whereas the matrix transpose is fast and accurate.

Do such matrices really exist in the wild, or are they mere figments of a data scientist's imagination? Yes, they really do exist. In fact, the identity matrix is an example of an orthogonal matrix. Here are another two:

$$\frac{1}{\sqrt{2}}\begin{bmatrix} 1 & -1 \\ 1 & 1 \end{bmatrix}, \quad \frac{1}{3}\begin{bmatrix} 1 & 2 & 2 \\ 2 & 1 & -2 \\ -2 & 2 & -1 \end{bmatrix}$$

Please take a moment to confirm that each column has unit length and is orthogonal to other columns. Then we can confirm in Python:

```
Q1 = np.array([ [1,-1],[1,1] ]) / np.sqrt(2)
Q2 = np.array([ [1,2,2],[2,1,-2],[-2,2,-1] ]) / 3

print( Q1.T @ Q1 )
print( Q2.T @ Q2 )
```

Both outputs are the identity matrix (within rounding errors on the order of 10^{-15}). What happens if you compute $\mathbf{Q}\mathbf{Q}^T$? Is that still the identity matrix? Try it to find out![1]

Another example of an orthogonal matrix is the pure rotation matrices that you learned about in Chapter 7. You can go back to that code and confirm that the

---

1 This is further explored in Exercise 9-1.

transformation matrix times its transpose is the identity matrix, regardless of the rotation angle (as long as the same rotation angle is used in all matrix elements). Permutation matrices are also orthogonal. Permutation matrices are used to exchange rows of a matrix; you'll learn about them in the discussion of LU decomposition in the next chapter.

How do you create such majestic marvels of mathematics? An orthogonal matrix can be computed from a nonorthogonal matrix via QR decomposition, which is basically a sophisticated version of Gram-Schmidt. And how does Gram-Schmidt work? That's basically the orthogonal vector decomposition that you learned about in Chapter 2.

Gram-Schmidt

The Gram-Schmidt procedure is a way to transform a nonorthogonal matrix into an orthogonal matrix. Gram-Schmidt has high educational value, but unfortunately very little application value. The reason is that—as you've now read several times before— there are numerical instabilities resulting from many divisions and multiplications by tiny numbers. Fortunately, there are more sophisticated and numerically stable methods for QR decomposition, such as Householder reflections. The details of that algorithm are outside the scope of this book, but they are handled by low-level numerical computation libraries that Python calls.

Nevertheless, I'm going to describe the Gram-Schmidt procedure (sometimes abbreviated to GS or G-S) because it shows an application of orthogonal vector decomposition, because you are going to program the algorithm in Python based on the following math and description, and because GS is the right way to conceptualize how and why QR decomposition works even if the low-level implementation is slightly different.

A matrix $\mathbf{V}$ comprising columns $\mathbf{v}_1$ through $\mathbf{v}_n$ is transformed into an orthogonal matrix $\mathbf{Q}$ with columns $\mathbf{q}_k$ according to the following algorithm.

For all column vectors in $\mathbf{V}$ starting from the first (leftmost) and moving systematically to the last (rightmost):

1. Orthogonalize $\mathbf{v}_k$ to all previous columns in matrix $\mathbf{Q}$ using orthogonal vector decomposition. That is, compute the component of $\mathbf{v}_k$ that is perpendicular to $\mathbf{q}_{k-1}, \mathbf{q}_{k-2}$, and so on down to $\mathbf{q}_1$. The orthogonalized vector is called $\mathbf{v}_k^*$.[2]

2. Normalize $\mathbf{v}_k^*$ to unit length. This is now $\mathbf{q}_k$, the kth column in matrix $\mathbf{Q}$.

---

2 The first column vector is not orthogonalized because there are no preceeding vectors; therefore, you begin with the following normalization step.

Sounds simple, doesn't it? Implementing this algorithm in code can be tricky because of the repeated orthogonalizations. But with some perseverence, you can figure it out (Exercise 9-2).

QR Decomposition

The GS procedure transforms a matrix into an orthogonal matrix $\mathbf{Q}$. (As I wrote in the previous section, $\mathbf{Q}$ is actually obtained using a series of vector-plane reflections known as the Householder transformation, but that's due to numerical issues; GS is a great way to conceptualize how $\mathbf{Q}$ matrices are formed.)

What's in a Sound?

The "QR" in QR decomposition is pronounced "queue are." In my opinion, that's a real missed opportunity; linear algebra would be more fun to learn if we pronounced it "QweRty decomposition." Or maybe we could have pronounced it "core decomposition" to appeal to the fitness crowd. But, for better and for worse, modern conventions are shaped by historical precedent.

$\mathbf{Q}$ is obviously different from the original matrix (assuming the original matrix was not orthogonal). So we have lost information about that matrix. Fortunately, that "lost" information can be easily retrieved and stored in another matrix $\mathbf{R}$ that multiplies $\mathbf{Q}$.[3] That leads to the question of how we create $\mathbf{R}$. In fact, creating $\mathbf{R}$ is straightforward and comes right from the definition of QR:

$$\mathbf{A} = \mathbf{QR}$$
$$\mathbf{Q}^T\mathbf{A} = \mathbf{Q}^T\mathbf{QR}$$
$$\mathbf{Q}^T\mathbf{A} = \mathbf{R}$$

Here you see the beauty of orthogonal matrices: we can solve matrix equations without having to worry about computing the inverse.

The following Python code shows how to compute QR decomposition of a square matrix, and Figure 9-1 illustrates the three matrices:

```
A = np.random.randn(6,6)
Q,R = np.linalg.qr(A)
```

---

3 Recovering $\mathbf{R}$ through matrix multiplication is possible because GS is a series of linear transformations.

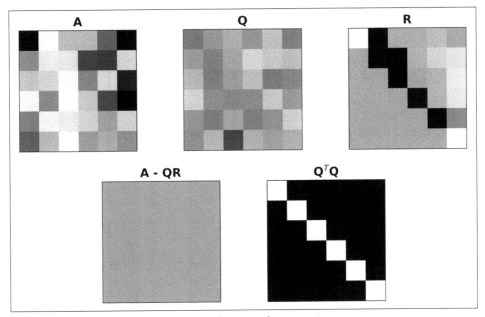

Figure 9-1. QR decomposition of a random-numbers matrix

Several important features of QR decomposition are visible in Figure 9-1, including that $\mathbf{A} = \mathbf{QR}$ (their difference is the zeros matrix) and that $\mathbf{Q}$ times its transpose gives the identity matrix.

Check out the $\mathbf{R}$ matrix: it's upper-triangular (all elements below the diagonal are zero). That seems unlikely to have occurred by chance, considering we started from a random matrix. In fact, the $\mathbf{R}$ matrix is *always* upper-triangular. To understand why, you need to think about the GS algorithm and the organization of the dot products in matrix multiplication. I will explain why $\mathbf{R}$ is upper-triangular in the next section; before then, I want you to come up with an answer.

Sizes of Q and R

The sizes of $\mathbf{Q}$ and $\mathbf{R}$ depend on the size of the to-be-decomposed matrix $\mathbf{A}$ and on whether the QR decomposition is "economy" (also called "reduced") or "full" (also called "complete"). Figure 9-2 shows an overview of all possible sizes.

Economy versus full QR decomposition applies only to tall matrices. The question is this: for a tall matrix ($M > N$), do we create a $\mathbf{Q}$ matrix with N columns or M columns? The former option is called *economy* or *reduced*, and gives a tall $\mathbf{Q}$; and the latter option is called *full* or *complete*, and gives a square $\mathbf{Q}$.

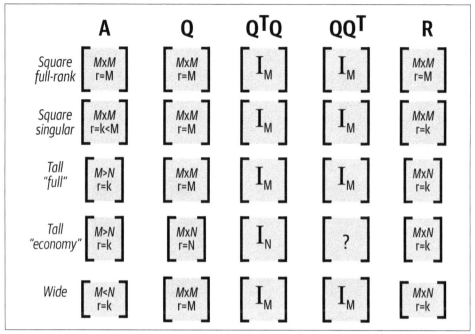

*Figure 9-2. Sizes of **Q** and **R** depending on the size of **A**. The "?" indicates that the matrix elements depend on the values in **A**, i.e., it is not the identity matrix.*

It may seem surprising that **Q** can be square when **A** is tall (in other words, that **Q** can have more columns than **A**): where do the extra columns come from? It is, in fact, possible to craft orthogonal vectors "out of thin air." Consider the following example in Python:

```python
A = np.array([ [1,-1] ]).T
Q,R = np.linalg.qr(A,'complete')
Q*np.sqrt(2) # scaled by sqrt(2) to get integers

>> array([[-1.,  1.],
          [ 1.,  1.]])
```

Notice the optional second input `'complete'`, which produces a full QR decomposition. Setting that to `'reduced'`, which is the default, gives the economy-mode QR decomposition, in which **Q** is the same size as **A**.

Because it is possible to craft more than $M > N$ orthogonal vectors from a matrix with N columns, the rank of **Q** is always the maximum possible rank, which is M for all square **Q** matrices and N for the economy **Q**. The rank of **R** is the same as the rank of **A**.

The difference in rank between **Q** and **A** resulting from orthogonalization means that **Q** spans all of $\mathbb{R}^M$ even if the column space of **A** is only a lower-dimensional subspace of $\mathbb{R}^M$. That fact is central to why the singular value decomposition is so useful for revealing properties of a matrix, including its rank and null space. Yet another reason to look forward to learning about the SVD in Chapter 14!

A note about uniqueness: QR decomposition is not unique for all matrix sizes and ranks. This means that it is possible to obtain $\mathbf{A} = \mathbf{Q}_1\mathbf{R}_1$ and $\mathbf{A} = \mathbf{Q}_2\mathbf{R}_2$ where $\mathbf{Q}_1 \neq \mathbf{Q}_2$. However, all QR decomposition results have the same properties described in this section. QR decomposition can be made unique given additional constraints (e.g., positive values on the diagonals of **R**), although this is not necessary in most cases, and is not implemented in Python or MATLAB. You'll see this nonuniqueness when comparing GS to QR in Exercise 9-2.

Why R is upper triangular

I hope you gave this question some serious thought. It's a tricky point about QR decomposition, so if you couldn't figure it out on your own, then please read the following few paragraphs, and then look away from the book/screen and rederive the argument.

I will start by reminding you of three facts:

- **R** comes from the formula $\mathbf{Q}^T\mathbf{A} = \mathbf{R}$.
- The lower triangle of a product matrix comprises dot products between *later* rows of the left matrix and *earlier* columns of the right matrix.
- The rows of $\mathbf{Q}^T$ are the columns of **Q**.

Putting those together: because orthogonalization works column-wise from left to right, *later* columns in **Q** are orthogonalized to *earlier* columns of **A**. Therefore, the lower triangle of **R** comes from pairs of vectors that have been orthogonalized. In contrast, *earlier* columns in **Q** are not orthogonalized to *later* columns of **A**, so we would not expect their dot products to be zero.

Final comment: if columns i and j of **A** were already orthogonal, then the corresponding (i,j)th element in **R** would be zero. In fact, if you compute the QR decomposition of an orthogonal matrix, then **R** will be a diagonal matrix in which the diagonal elements are the norms of each column in **A**. That means that if $\mathbf{A} = \mathbf{Q}$, then $\mathbf{R} = \mathbf{I}$, which is obvious from the equation solved for **R**. You'll get to explore this in Exercise 9-3.

QR and Inverses

QR decomposition provides a more numerically stable way to compute the matrix inverse.

Let's start by writing out the QR decomposition formula and inverting both sides of the equation (note the application of the LIVE EVIL rule):

$$\mathbf{A} = \mathbf{QR}$$
$$\mathbf{A}^{-1} = (\mathbf{QR})^{-1}$$
$$\mathbf{A}^{-1} = \mathbf{R}^{-1}\mathbf{Q}^{-1}$$
$$\mathbf{A}^{-1} = \mathbf{R}^{-1}\mathbf{Q}^{\mathrm{T}}$$

Thus, we can obtain the inverse of $\mathbf{A}$ as the inverse of $\mathbf{R}$ times the transpose of $\mathbf{Q}$. $\mathbf{Q}$ is numerically stable due to the Householder reflection algorithm, and $\mathbf{R}$ is numerically stable because it simply results from matrix multiplication.

Now, we still need to invert $\mathbf{R}$ explicitly, but inverting triangular matrices is highly numerically stable when done through a procedure called back substitution. You'll learn more about that in the next chapter, but the key point is this: an important application of QR decomposition is providing a more numerically stable way to invert matrices, compared to the algorithm presented in the previous chapter.

On the other hand, keep in mind that matrices that are theoretically invertible but are close to singular are still very difficult to invert; QR decomposition may be *more* numerically stable than the traditional algorithm presented in the previous chapter, but that doesn't guarantee a high-quality inverse. A rotten apple dipped in honey is still rotten.

Summary

QR decomposition is great. It's definitely on my list of the top five awesomest matrix decompositions in linear algebra. Here are the key take-home messages of this chapter:

- An orthogonal matrix has columns that are pair-wise orthogonal and with norm = 1. Orthogonal matrices are key to several matrix decompositions, including QR, eigendecomposition, and singular value decomposition. Orthogonal matrices are also important in geometry and computer graphics (e.g., pure rotation matrices).

- You can transform a nonorthogonal matrix into an orthogonal matrix via the Gram-Schmidt procedure, which involves applying orthogonal vector decomposition to isolate the component of each column that is orthogonal to all previous columns ("previous" meaning left to right).

- QR decomposition is the result of Gram-Schmidt (technically, it is implemented by a more stable algorithm, but GS is still the right way to understand it).

Code Exercises

Exercise 9-1.

A square **Q** has the following equalities:

$$\mathbf{Q}^\mathrm{T}\mathbf{Q} = \mathbf{Q}\mathbf{Q}^\mathrm{T} = \mathbf{Q}^{-1}\mathbf{Q} = \mathbf{Q}\mathbf{Q}^{-1} = \mathbf{I}$$

Demonstrate this in code by computing **Q** from a random-numbers matrix, then compute $\mathbf{Q}^\mathrm{T}$ and $\mathbf{Q}^{-1}$. Then show that all four expressions produce the identity matrix.

Exercise 9-2.

Implement the Gram-Schmidt procedure as described earlier.[4] Use a 4×4 random-numbers matrix. Check your answer against **Q** from np.linalg.qr.

Important: there is a fundamental sign uncertainty in transformations like the Householder reflection. This means that vectors can "flip" (be multiplied by −1) depending on minor differences in algorithm and implementation. This feature exists in many matrix decompositions including eigenvectors. I have a longer and more in-depth discussion of why this is and what it means in Chapter 13. For now, the upshot is this: *subtract* your **Q** from Python's **Q** and *add* your **Q** and Python's **Q**. Nonzero columns in one will be zeros in the other.

Exercise 9-3.

In this exercise, you will find out what happens when you apply QR decomposition to a matrix that is almost-but-not-quite orthogonal. First, create an orthogonal matrix, called **U**, from the QR decomposition of a 6×6 random-numbers matrix. Compute the QR decomposition of **U**, and confirm that $\mathbf{R} = \mathbf{I}$ (and make sure you understand why!).

Second, modify the norms of each column of **U**. Set the norms of columns 1–6 to be 10–15 (that is, the first column of **U** should have a norm of 10, the second column

---

4 Take your time with this exercise; it's quite challenging.

should have a norm of 11, and so on). Run that modulated $\mathbf{U}$ matrix through QR decomposition and confirm that its $\mathbf{R}$ is a diagonal matrix with diagonal elements equaling 10–15. What is $\mathbf{Q}^T\mathbf{Q}$ for this matrix?

Third, break the orthogonality of $\mathbf{U}$ by setting element $u_{1,4} = 0$. What happens to $\mathbf{R}$ and why?

Exercise 9-4.

The purpose of this exercise is to compare the numerical errors using the "old-school" inverse method you learned in the previous chapter to the QR-based inverse method. We will use random-numbers matrices, keeping in mind that they tend to be numerically stable and thus have accurate inverses.

Here's what to do: copy the code from Exercise 8-2 into a Python function that takes a matrix as input and provides its inverse as output. (You can also include a check that the input matrix is square and full-rank.) I called this function oldSchoolInv. Next, create a 5×5 random-numbers matrix. Compute its inverse using the old-school method and the QR decomposition method introduced in this chapter (you can use your "old-school method" to compute $\mathbf{R}^{-1}$). Compute the inverse-estimation error as the Euclidean distance from the matrix times its inverse to the true identity matrix from np.eye. Make a barplot of the results, showing the two methods on the x-axis and the error (Euclidean distance to $\mathbf{I}$) on the y-axis, as in Figure 9-3.

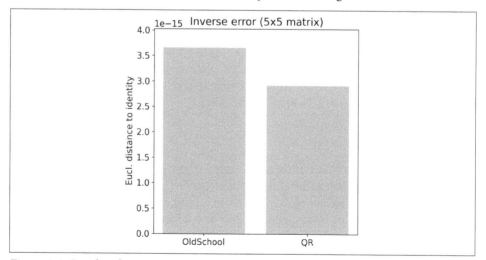

Figure 9-3. Results of Exercise 9-4

Run the code multiple times and inspect the barplot. You'll find that sometimes the old-school method is better while other times the QR method is better (smaller numbers are better; in theory, the bars should have zero height). Try it again using a 30 × 30 matrix. Are the results more consistent? In fact, there is a lot of variance from run to run. This means we should run an experiment where we repeat the comparison many times. That's the next exercise.

Exercise 9-5.

Put the code from the previous exercise into a for loop over a hundred iterations in which you repeat the experiment, each time using a different random-numbers matrix. Store the error (Euclidean distance) for each iteration, and make a plot like Figure 9-4, which shows the average over all experiment runs (gray bar) and all individual errors (black dots). Run the experiment for 5 × 5 and 30 × 30 matrices.

You can also try using np.linalg.inv to invert **R** instead of the old-school method to see if that has an effect.

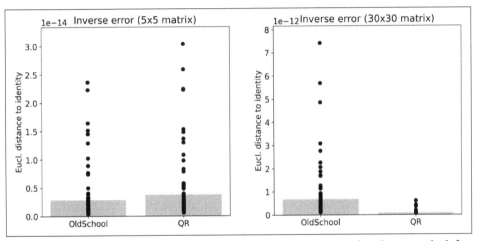

Figure 9-4. Results of Exercise 9-5. Note the difference in y-axis scaling between the left and right panels.

Exercise 9-6.

An interesting property of square orthogonal matrices is that all of their singular values (and their eigenvalues) are 1. That means that they have an induced 2-norm of 1 (the induced norm is the largest singular value), and they have a Frobenius norm of M. The latter result is because the Frobenius norm equals the square root of the sum of the squared singular values. In this exercise, you will confirm these properties.

Create an $M \times M$ orthogonal matrix as the QR decomposition of a random matrix. Compute its induced 2-norm using np.linalg.norm and compute its Frobenius norm using the equation you learned in Chapter 6, divided by the square root of

M. Confirm that both quantities are 1 (to within a reasonable tolerance of rounding error). Check using several different values of *M*.

Next, explore the meaning of the induced norm using matrix-vector multiplication. Create a random *M*-element column vector $\mathbf{v}$. Then compute the norms of $\mathbf{v}$ and $\mathbf{Qv}$. Those norms should equal each other (although you wouldn't expect them to equal 1).

Finally, get a piece of paper and develop a proof of that empirical demonstration. That proof is printed in the next paragraph, so don't look down! But you can check the footnote if you need a hint.[5]

I sincerely hope you are reading this to check your reasoning, not because you are cheating! Anyway, the proof is that the vector norm $\| \mathbf{v} \|$ can be computed as $\mathbf{v}^T\mathbf{v}$; therefore, the vector norm $\| \mathbf{Qv} \|$ is computed as $(\mathbf{Qv})^T\mathbf{Qv} = \mathbf{v}^T\mathbf{Q}^T\mathbf{Qv}$. The $\mathbf{Q}^T\mathbf{Q}$ cancels to give the identity matrix, leaving the dot product of the vector with itself. The conclusion is that orthogonal matrices can rotate but never scale a vector.

Exercise 9-7.

This exercise will highlight one feature of the $\mathbf{R}$ matrix that is relevant for understanding how to use QR to implement least squares (Chapter 12): when $\mathbf{A}$ is tall and full column-rank, the first *N* rows of $\mathbf{R}$ are upper-triangular, whereas rows *N* + 1 through *M* are zeros. Confirm this in Python using a random 10×4 matrix. Make sure to use the complete (full) QR decomposition, not the economy (compact) decomposition.

Of course, $\mathbf{R}$ is noninvertible because it is nonsquare. But (1) the submatrix comprising the first *N* rows is square and full-rank (when $\mathbf{A}$ is full column-rank) and thus has a full inverse, and (2) the tall $\mathbf{R}$ has a pseudoinverse. Compute both inverses, and confirm that the full inverse of the first *N* rows of $\mathbf{R}$ equals the first *N* columns of the pseudoinverse of the tall $\mathbf{R}$.

---

5 Hint: write down the dot-product formula for the vector norm.

## CHAPTER 10

Row Reduction and LU Decomposition

Now we move to LU decomposition. LU, like QR, is one of the computational backbones underlying data-science algorithms, including least squares model fitting and the matrix inverse. This chapter is, therefore, pivotal to your linear algebra education.

The thing about LU decomposition is you cannot simply learn it immediately. Instead, you first need to learn about systems of equations, row reduction, and Gaussian elimination. And in the course of learning those topics, you'll also learn about echelon matrices and permutation matrices. Oh yes, dear reader, this will be an exciting and action-packed chapter.

Systems of Equations

To understand LU decomposition and its applications, you need to understand row reduction and Gaussian elimination. And to understand those topics, you need to understand how to manipulate equations, convert them into a matrix equation, and solve that matrix equation using row reduction.

Let's start with a "system" of one equation:

$$2x = 8$$

As I'm sure you learned in school, you can do various mathematical manipulations to the equation—as long as you do the same thing to both sides of the equation. That means that the following equation is not the same as the previous one, but they are related to each other by simple manipulations. More importantly, any solution to one equation is a solution to the other:

$$5(2x - 3) = 5(8 - 3)$$

Now let's move to a system of two equations:

$$x = 4 - y$$
$$y = x/2 + 2$$

In this system of equations, it is impossible to solve for unique values of x and y from either of those equations alone. Instead, you need to consider both equations simultaneously to derive the solution. If you try to solve that system now, you would probably take the strategy of substituting y in the first equation with the right-hand side of the second equation. After solving for x in the first equation, you plug that value into the second equation to solve for y. This strategy is similar to (though not as efficient as) back substitution, which I'll define later.

An important feature of a system of equations is that you can add or subtract individual equations to each other. In the following equations, I've added two times the second equation to the first, and subtracted the first original equation from the second (parentheses added for clarity):

$$x + (2y) = 4 - y + (x + 4)$$
$$y - (x) = x/2 + 2 - (4 - y)$$

I will let you work through the arithmetic, but the upshot is that x drops out of the first equation while y drops out of the second equation. That makes the solution much easier to calculate ($x = 4/3$, $y = 8/3$). Here's the important point: scalar multiplying equations and adding them to other equations made the solution to the system easier to find. Again, the modulated and original systems are not the same equations, but their solutions are the same because the two systems are linked by a series of linear operations.

This is the background knowledge you need to learn how to solve systems of equations using linear algebra. But before learning that approach, you need to learn how to represet a system of equations using matrices and vectors.

Converting Equations into Matrices

Converting systems of equations into a matrix-vector equation is used to solve systems of equations, and it's used to set up the formula for the general linear model in statistics. Fortunately, translating equations into matrices is conceptually simple, and involves two steps.

First, organize the equations so that the constants are on the right-hand side of the equations. The *constants* are the numbers that are unattached to the variables (sometimes called *intercepts* or *offsets*). The variables and their multiplying coefficients are on the left-hand side of the equation, in the same order (e.g., all equations should have the x term first, then the y term, and so on). The following equations form the system of equations we've been working with, in the proper organization:

$$x + y = 4$$
$$-x/2 + y = 2$$

Second, separate the coefficients (the numbers multiplying the variables; variables that are missing from an equation have a coefficient of zero) into a matrix with one row per equation. The variables are placed into a column vector that right-multiplies the coefficients matrix. And the constants are placed into a column vector on the right-hand side of the equation. Our example system has a matrix equation that looks like this:

$$\begin{bmatrix} 1 & 1 \\ -1/2 & 1 \end{bmatrix} \begin{bmatrix} x \\ y \end{bmatrix} = \begin{bmatrix} 4 \\ 2 \end{bmatrix}$$

And voilà! You've converted a system of equations into one matrix equation. We can refer to this equation as $\mathbf{Ax} = \mathbf{b}$, where $\mathbf{A}$ is the matrix of coefficients, $\mathbf{x}$ is a vector of unknown variables to solve for (in this case, $\mathbf{x}$ is the vector comprising $[x\,y]$), and $\mathbf{b}$ is a vector of constants.

Please take a moment to make sure you understand how the matrix equation maps onto the system of equations. In particular, work through the matrix-vector multiplication to demonstrate that it equals the original system of equations.

Working with Matrix Equations

You can manipulate matrix equations just like normal equations, including adding, multiplying, transposing, etc., as long as the manipulations are valid (e.g., matrix sizes match for addition) and all manipulations affect both sides of the equation. For example, the following progression of equations is valid:

$$\mathbf{Ax} = \mathbf{b}$$
$$\mathbf{v} + \mathbf{Ax} = \mathbf{v} + \mathbf{b}$$
$$(\mathbf{v} + \mathbf{Ax})^{\mathrm{T}} = (\mathbf{v} + \mathbf{b})^{\mathrm{T}}$$

The main difference between working with matrix equations versus scalar equations is that because matrix multiplication is side-dependent, you must multiply matrices in the same way on both sides of the equation.

For example, the following progression of equations is valid:

$$\mathbf{AX} = \mathbf{B}$$
$$\mathbf{CAX} = \mathbf{CB}$$

Notice that **C** premultiplies both sides of the equation. In contrast, the following progression is **not** valid:

$$\mathbf{AX} = \mathbf{B}$$
$$\mathbf{AXC} = \mathbf{CB}$$

The problem here is that **C** *post*multiplies in the left-hand side but *pre*multiplies in the right-hand side. To be sure, there will be a few exceptional cases where that equation is valid (e.g., if **C** is the identity or zeros matrix), but in general that progression is not valid.

Let's see an example in Python. We will solve for the unknown matrix **X** in the equation **AX** = **B**. The following code generates **A** and **B** from random numbers. You already know that we can solve for **X** by using $\mathbf{A}^{-1}$. The question is whether the order of multiplication matters.[1]

```
A = np.random.randn(4,4)
B = np.random.randn(4,4)

# solve for X
X1 = np.linalg.inv(A) @ B
X2 = B @ np.linalg.inv(A)

# residual (should be zeros matrix)
res1 = A@X1 - B
res2 = A@X2 - B
```

If matrix multiplication were commutative (meaning that the order doesn't matter), then res1 and res2 should both equal the zeros matrix. Let's see:

```
res1:

[[-0.  0.  0.  0.]
 [-0. -0.  0.  0.]
 [ 0.  0.  0.  0.]
 [ 0.  0. -0. -0.]]

res2:
```

---

1 Of course you know that order matters, but empirical demonstrations help build intuition. And I want you to get in the habit of using Python as a tool to empirically confirm principles in mathematics.

```
[[-0.47851507  6.24882633  4.39977191  1.80312482]
 [ 2.47389146  2.56857366  1.58116135 -0.52646367]
 [-2.12244448 -0.20807188  0.2824044  -0.91822892]
 [-3.61085707 -3.80132548 -3.47900644 -2.372463  ]]
```

Now you know how to express a system of equations using one matrix equation. I'm going to get back to this in a few sections; first, I need to teach you about row reduction and the echelon form of a matrix.

Row Reduction

Row reduction is a topic that gets a lot of attention in traditional linear algebra, because it is the time-honored way to solve systems of equations *by hand*. I seriously doubt you'll solve any systems of equations by hand in your career as a data scientist. But row reduction is useful to know about, and it leads directly to LU decomposition, which actually is used in applied linear algebra. So let's begin.

Row reduction means iteratively applying two operations—scalar multiplication and addition—to rows of a matrix. Row reduction relies on the same principle as adding equations to other equations within a system.

Memorize this statement: *The goal of row reduction is to transform a dense matrix into an upper-triangular matrix.*

Let's start with a simple example. In the following dense matrix, we add the first row to the second row, which knocks out the −2. And with that, we've transformed our dense matrix into an upper-triangular matrix:

$$\begin{bmatrix} 2 & 3 \\ -2 & 2 \end{bmatrix} \xrightarrow{R_1 + R_2} \begin{bmatrix} 2 & 3 \\ 0 & 5 \end{bmatrix}$$

The upper-triangular matrix that results from row reduction is called the *echelon form* of the matrix.

Formally, a matrix is in echelon form if (1) the leftmost nonzero number in a row (which is called the *pivot*) is to the right of the pivot of rows above, and (2) any rows of all zeros are below rows containing nonzeros.

Similar to manipulating equations in a system, the matrix *after* row reduction is different from the matrix *before* row reduction. But the two matrices are linked by a linear transform. And because linear transforms can be represented by matrices, we can use matrix multiplication to express row reduction:

$$\begin{bmatrix} 1 & 0 \\ 1 & 1 \end{bmatrix} \begin{bmatrix} 2 & 3 \\ -2 & 2 \end{bmatrix} = \begin{bmatrix} 2 & 3 \\ 0 & 5 \end{bmatrix}$$

I will call that matrix $\mathbf{L}^{-1}$ for reasons that will become clear when I introduce LU decomposition.[2] Thus, in the expression $\left(\mathbf{L}^{-1}\mathbf{A} = \mathbf{U}\right)$, $\mathbf{L}^{-1}$ is the linear transformation that keeps track of the manipulations we've implemented through row reduction. For now, you don't need to focus on $\mathbf{L}^{-1}$—in fact, it's often ignored during Gaussian elimination. But the key point (slightly expanded from an earlier claim) is this: *row reduction involves transforming a matrix into an upper-triangular matrix via row manipulations, which can be implemented as premultiplication by a transformation matrix.*

Here's another example of a 3×3 matrix. This matrix requires two steps to transform into its echelon form:

$$\begin{bmatrix} 1 & 2 & 2 \\ -1 & 3 & 0 \\ 2 & 4 & -3 \end{bmatrix} \xrightarrow{-2R_1 + R_3} \begin{bmatrix} 1 & 2 & 2 \\ -1 & 3 & 0 \\ 0 & 0 & -7 \end{bmatrix} \xrightarrow{R_1 + R_2} \begin{bmatrix} 1 & 2 & 2 \\ 0 & 5 & 2 \\ 0 & 0 & -7 \end{bmatrix}$$

Row reduction is tedious (see "Is Row Reduction Always So Easy?"). Surely there must be a Python function to do it for us! Well, there is and there isn't. There is no Python function that returns an echelon form like what I created in the two previous examples. The reason is that the echelon form of a matrix is not unique. For example, in the previous 3×3 matrix, you could multiply the second row by 2 to give a row vector of [0 10 4]. That creates a perfectly valid—but different—echelon form of the same original matrix. Indeed, there is an infinite number of echelon matrices associated with that matrix.

That said, two echelon forms of a matrix are preferred over the infinite possible echelon forms. Those two forms are unique given some constraints and are called the reduced row echelon form and $\mathbf{U}$ from LU decomposition. I will introduce both later; first, it's time to learn how to use row reduction to solve systems of equations.

Is Row Reduction Always So Easy?

Row reducing a matrix to its echelon form is a skill that requires diligent practice and learning several tricks with cool-sounding names like row swapping and partial pivoting. Row reduction reveals several interesting properties of the row and column spaces of the matrix. And although successfully row reducing a 5×6 matrix with row swaps by hand brings a feeling of accomplishment and satisfaction, my opinion is that your time is better spent on concepts in linear algebra that have more direct applica-

---

2 Spoiler alert: LU decomposition involves representing a matrix as the product of a lower-triangular and an upper-triangular matrix.

tion to data science. We are building toward understanding LU decomposition, and this brief introduction to row reduction is sufficient for that goal.

Gaussian Elimination

At this point in the book, you know how to solve matrix equations using the matrix inverse. What if I told you that you could solve a matrix equation without inverting any matrices?[3]

This technique is called *Gaussian elimination*. Despite its name, the algorithm was actually developed by Chinese mathematicians nearly two thousand years before Gauss and then rediscovered by Newton hundreds of years before Gauss. But Gauss made important contributions to the method, including the techniques that modern computers implement.

Gaussian elimination is simple: augment the matrix of coefficients by the vector of constants, row reduce to echelon form, and then use back substitution to solve for each variable in turn.

Let's start with the system of two equations that we solved earlier:

$$x = 4 - y$$
$$y = x/2 + 2$$

The first step is to convert this system of equations into a matrix equation. We've already worked through this step; that equation is printed here as a reminder:

$$\begin{bmatrix} 1 & 1 \\ -1/2 & 1 \end{bmatrix} \begin{bmatrix} x \\ y \end{bmatrix} = \begin{bmatrix} 4 \\ 2 \end{bmatrix}$$

Next, we augment the coefficients matrix with the constants vector:

$$\begin{bmatrix} 1 & 1 & 4 \\ -1/2 & 1 & 2 \end{bmatrix}$$

Then we row reduce that augmented matrix. Note that the column vector of constants will change during row reduction:

---

3 Please conjure into your imagination that *Matrix* meme with Morpheus proffering the red and blue pill, corresponding to new knowledge versus sticking to what you know.

$$\begin{bmatrix} 1 & 1 & 4 \\ -1/2 & 1 & 2 \end{bmatrix} \xrightarrow{\ 1/2R_1 + R_2\ } \begin{bmatrix} 1 & 1 & 4 \\ 0 & 3/2 & 4 \end{bmatrix}$$

Once we have the matrix in its echelon form, we translate the augmented matrix back into a system of equations. That looks like this:

$$\begin{aligned} x + \quad y &= 4 \\ 3/2y &= 4 \end{aligned}$$

Gaussian elimination via row reduction removed the x term in the second equation, which means that solving for y merely involves some arithmetic. Once you solve for $y = 8/3$, plug that value into y in the first equation and solve for x. This procedure is called *back substitution*.

In the previous section, I wrote that Python does not have a function to compute the echelon form of a matrix because it is not unique. And then I wrote that there is one unique echelon matrix, called the *reduced row echelon form* and often abbreviated RREF, which Python will compute. Keep reading to learn more…

Gauss-Jordan Elimination

Let's keep row reducing our example matrix with the goal of turning all the *pivots* —the leftmost nonzero numbers in each row—into 1s. Once you have the echelon matrix, you simply divide each row by its pivot. In this example, the first row already has a 1 in the leftmost position, so we just need to adjust the second row. That gives us the following matrix:

$$\begin{bmatrix} 1 & 1 & 4 \\ 0 & 1 & 8/3 \end{bmatrix}$$

And now for the trick: we continue row reducing *upward* to eliminate all the elements above each pivot. In other words, we want an echelon matrix in which each pivot is 1 and it's the only nonzero number in its column.

$$\begin{bmatrix} 1 & 1 & 4 \\ 0 & 1 & 8/3 \end{bmatrix} \xrightarrow{\ -R_2 + R_1\ } \begin{bmatrix} 1 & 0 & 4/3 \\ 0 & 1 & 8/3 \end{bmatrix}$$

That's the RREF of our original matrix. You can see the identity matrix on the left—RREF will always produce an identity matrix as a submatrix in the upper-left of the original matrix. That's a result of setting all pivots to 1 and using upward row reduction to eliminate all elements above each pivot.

Now we continue with Gaussian elimination by translating the matrix back into a system of equations:

$$x = 4/3$$
$$y = 8/3$$

We no longer need back substitution, or even basic arithmetic: the modified Gaussian elimination—which is called Gauss-Jordan elimination—decoupled the interwoven variables in the system of equations, and laid bare the solutions to each variable.

Gauss-Jordan elimination was how people solved systems of equations by hand for over a century before computers came along to help us with the number crunching. In fact, computers still implement this exact same method, with only a few small modifications to ensure numerical stability.

The RREF is unique, meaning that a matrix has exactly one associated RREF. NumPy does not have a function to compute the RREF of a matrix, but the sympy library does (sympy is the symbolic math library in Python and is a powerful engine for "chalkboard math"):

```
import sympy as sym

# the matrix converted to sympy
M = np.array([ [1,1,4],[-1/2,1,2] ])
symMat = sym.Matrix(M)

# RREF
symMat.rref()[0]

>>
  [[1, 0, 1.33333333333333],
   [0, 1, 2.66666666666667]]
```

Matrix Inverse via Gauss-Jordan Elimination

The key insight from Gauss-Jordan elimination is that row reduction produces a sequence of row manipulations that solves a set of equations. Those row manipulations are linear transformations.

Curiously, the description of Gauss-Jordan elimination is consistent with the description of the matrix inverse: a linear transformation that solves a set of equations. But wait, what "system of equations" does the matrix inverse solve? A fresh perspective on the matrix inverse will provide some new insights. Consider this system of equations:

$$ax_1 + by_1 = 1$$
$$cx_1 + dy_1 = 0$$

Translated into a matrix equation, we get:

$$\begin{bmatrix} a & b \\ c & d \end{bmatrix} \begin{bmatrix} x_1 \\ y_1 \end{bmatrix} = \begin{bmatrix} 1 \\ 0 \end{bmatrix}$$

Check out that constants vector—it is the first column of the 2×2 identity matrix! That means that applying RREF to a square full-rank matrix augmented by the first column of the identity matrix will reveal the linear transformation that brings the matrix into the first column of the identity matrix. And that in turn means that the vector $[x_1 \ y_1]^T$ is the first column of the matrix inverse.

We then repeat the procedure but solving for the second column of the inverse:

$$ax_2 + by_2 = 0$$
$$cx_2 + dy_2 = 1$$

RREF on that system gives the vector $[x_2 \ y_2]^T$, which is the second column of the matrix inverse.

I've separated the columns of the identity matrix to link back to the perspective of solving systems of equations. But we can augment the entire identity matrix and solve for the inverse with one RREF.

Here is the bird's-eye view of obtaining the matrix inverse via Gauss-Jordan elimination (the square brackets indicate augmented matrices with the vertical line separating the two constituent matrices):

$$rref([\mathbf{A} \mid \mathbf{I}]) \Rightarrow [\mathbf{I} \mid \mathbf{A}^{-1}]$$

This is interesting because it provides a mechanism for computing the matrix inverse without computing determinants. On the other hand, row reduction does involve a lot of division, which increases the risk of numerical precision errors. For example, imagine we have two numbers that are essentially zero plus rounding error. If we end up dividing those numbers during RREF, we could get the fraction $10^{-15}/10^{-16}$, which is really 0/0 but the answer will be 10.

The conclusion here is similar to what I discussed in the previous chapter about using QR to compute the matrix inverse: using Gauss-Jordan elimination to compute the matrix inverse is likely to be more numerically stable than the full algorithm for the inverse, but a matrix that is close to singular or that has a high condition number is difficult to invert, regardless of the algorithm used.

LU Decomposition

The "LU" in *LU decomposition* stands for "lower upper," as in lower-triangular, upper-triangular. The idea is to decompose a matrix into the product of two triangular matrices:

A = **LU**

Here's a numerical example:

$$\begin{bmatrix} 2 & 2 & 4 \\ 1 & 0 & 3 \\ 2 & 1 & 2 \end{bmatrix} = \begin{bmatrix} 1 & 0 & 0 \\ 1/2 & 1 & 0 \\ 1 & 1 & 1 \end{bmatrix} \begin{bmatrix} 2 & 2 & 4 \\ 0 & -1 & 1 \\ 0 & 0 & -3 \end{bmatrix}$$

And here's the corresponding Python code (note that the function for LU decomposition is in the SciPy library):

```
import scipy.linalg # LU in scipy library
A = np.array([ [2,2,4], [1,0,3], [2,1,2] ])
_,L,U = scipy.linalg.lu(A)

# print them out
print('L: '), print(L)
print('U: '), print(U)

L:
[[1.  0.  0. ]
 [0.5 1.  0. ]
 [1.  1.  1. ]]

U:
[[ 2.  2.  4.]
 [ 0. -1.  1.]
 [ 0.  0. -3.]]
```

Where do these two matrices come from? In fact, you already know the answer: row reduction can be expressed as $\mathbf{L}^{-1}\mathbf{A} = \mathbf{U}$, where $\mathbf{L}^{-1}$ contains the set of row manipulations that transforms the dense **A** into upper-triangular (echelon) **U**.

Because the echelon form is not unique, LU decomposition is not necessarily unique. That is, there is an infinite pairing of lower- and upper-triangular matrices that could multiply to produce matrix **A**. However, adding the constraint that the diagonals of **L** equal 1 ensures that LU decomposition is unique for a full-rank square matrix **A** (you can see this in the previous example). The uniqueness of LU decompositions of reduced-rank and nonsquare matrices is a longer discussion that I will omit here; however, SciPy's LU decomposition algorithm is determinstic, meaning that repeated LU decompositions of a given matrix will be identical.

Row Swaps via Permutation Matrices

Some matrices do not easily transform into an upper-triangular form. Consider the following matrix:

$$\begin{bmatrix} 3 & 2 & 1 \\ 0 & 0 & 5 \\ 0 & 7 & 2 \end{bmatrix}$$

It's not in echelon form, but it would be if we swapped the second and third rows. Row swaps are one of the tricks of row reduction, and are implemented through a permutation matrix:

$$\begin{bmatrix} 1 & 0 & 0 \\ 0 & 0 & 1 \\ 0 & 1 & 0 \end{bmatrix} \begin{bmatrix} 3 & 2 & 1 \\ 0 & 0 & 5 \\ 0 & 7 & 2 \end{bmatrix} = \begin{bmatrix} 3 & 2 & 1 \\ 0 & 7 & 2 \\ 0 & 0 & 5 \end{bmatrix}$$

Permutation matrices are often labeled **P**. Thus, the full LU decomposition actually takes the following form:

$$\mathbf{PA} = \mathbf{LU}$$

$$\mathbf{A} = \mathbf{P}^{\mathrm{T}}\mathbf{LU}$$

Remarkably, permutation matrices are orthogonal, and so $\mathbf{P}^{-1} = \mathbf{P}^{\mathrm{T}}$. Briefly, the reason is that all elements of a permutation matrix are either 0 or 1, and because rows are swapped only once, each column has exactly one nonzero element (indeed, all permutation matrices are identity matrices with row swaps). Therefore, the dot product of any two columns is 0 while the dot product of a column with itself is 1, meaning $\mathbf{P}^{\mathrm{T}}\mathbf{P} = \mathbf{I}$.

Important: the formulas I wrote above provide the *mathematical* description of LU decomposition. Scipy actually returns $\mathbf{A} = \mathbf{PLU}$, which we could also write as $\mathbf{P}^{\mathrm{T}}\mathbf{A} = \mathbf{LU}$. Exercise 10-4 provides the opportunity to explore this point of confusion.

Figure 10-1 shows an example of LU decomposition applied to a random matrix.

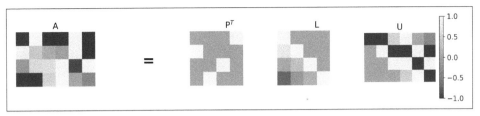

Figure 10-1. Visualization of LU decomposition

LU decomposition is used in several applications, including computing the determinant and the matrix inverse. In the next chapter, you'll see how LU decomposition is used in the least squares computation.

Summary

I opened this chapter by promising an action-packed educational adventure. I hope you experienced several adrenaline rushes while learning new perspectives on algebraic equations, matrix decompositions, and the matrix inverse. Here are the key take-home messages of this chapter:

- Systems of equations can be translated into a matrix equation. In addition to providing a compact representation, this allows for sophisticated linear algebra solutions to solving systems of equations.

- When working with matrix equations, remember that manipulations must be applied to both sides of the equation, and remember that matrix multiplication is noncommutative.

- Row reduction is a procedure in which the rows of matrix $\mathbf{A}$ are scalar multiplied and added, until the matrix is linearly transformed into an upper-triangular matrix $\mathbf{U}$. The set of linear transformations can be stored in another matrix $\mathbf{L}^{-1}$ that left-multiplies $\mathbf{A}$ to produce the expression $\mathbf{L}^{-1}\mathbf{A} = \mathbf{U}$.

- Row reduction has been used for centuries to solve systems of equations, including the matrix inverse, by hand. We still use row reduction, although computers take care of the arithmetic.

- Row reduction is also used to implement LU decomposition. LU decomposition is unique under some constraints, which are incorporated into the lu() function in SciPy.

Code Exercises

Exercise 10-1.

LU decomposition can be computationally intensive, although it is more efficient than other decompositions such as QR. Interestingly, LU decomposition is often used as a benchmark to compare computation times between operating systems, hardware processors, computer languages (e.g., C versus Python versus MATLAB), or implementation algorithms. Out of curiosity, I tested how long it took Python and MATLAB to run LU decomposition on a thousand matrices of size 100×100. On my laptop, MATLAB took around 300 ms while Python took around 410 ms. Python on Google Colab took around 1,000 ms. Test how long this takes on your computer.

Exercise 10-2.

Use the matrix-multiplication method to make a 6×8 rank-3 matrix. Take its LU decomposition, and show the three matrices with their ranks in the title, as in Figure 10-2. Notice the ranks of the three matrices and that **L** has all 1s on the diagonal. Feel free to explore the ranks of matrices with other sizes and ranks.

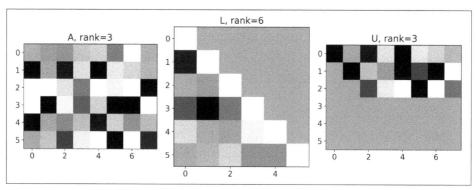

Figure 10-2. Results of Exercise 10-2

Exercise 10-3.

One application of LU decomposition is computing the determinant. Here are two properties of the determinant:[4] the determinant of a triangular matrix is the product of the diagonals, and the determinant of a product matrix equals the product of the determinants (that is, $det(\mathbf{AB}) = det(\mathbf{A})det(\mathbf{B})$). Putting these two facts together,

---

4 These are among the myriad aspects of linear algbera that you would learn in a traditional linear algebra textbook; they are interesting in their own right but less directly relevant to data science.

you can compute the determinant of a matrix as the product of the diagonals of **L** times the product of the diagonals of **U**. On the other hand, because the diagonals of **L** are all 1 (when implemented in Python to ensure uniqueness of the decomposition), then the determinant of a matrix **A** is simply the product of the diagonals of **U**. Try it in Python—and compare to the result of `np.linalg.det(A)`—multiple times with different random matrices, before reading the next paragraph.

Did you get the same result as Python? I assume you found that the determinants match in magnitude but that the signs would seemingly randomly differ. Why did that happen? It happened because I omitted the permutation matrix in the instructions. The determinant of a permutation matrix is +1 for an even number of row swaps and −1 for an odd number of row swaps. Now go back to your code and include the determinant of **P** in your computations.

Exercise 10-4.

Following the formula from the section "LU Decomposition" on page 169, the inverse of a matrix can be expressed as:

$$\mathbf{A} = \mathbf{P}^\mathrm{T}\mathbf{LU}$$
$$\mathbf{A}^{-1} = \left(\mathbf{P}^\mathrm{T}\mathbf{LU}\right)^{-1}$$
$$\mathbf{A}^{-1} = \mathbf{U}^{-1}\mathbf{L}^{-1}\mathbf{P}$$

Implement that third equation directly using the output from `scipy.linalg.lu` on a 4 × 4 random-numbers matrix. Is $\mathbf{AA}^{-1}$ the identity matrix? Sometimes it is and sometimes it isn't, depending on **P**. This discrepancy occurs because of what I wrote about the output of `scipy.linalg.lu`. Adjust the code so that it follows SciPy's convention instead of the math convention.

Here is the take-home message from this exercise: the absence of error messages does not necessarily mean your code is correct. Please sanity-check your mathematical code as much as possible.

Exercise 10-5.

For matrix $\mathbf{A} = \mathbf{PLU}$ (using Python's ordering of the permutation matrix), $\mathbf{A}^\mathrm{T}\mathbf{A}$ can be computed as $\mathbf{U}^\mathrm{T}\mathbf{L}^\mathrm{T}\mathbf{LU}$—*without the permutation matrices*. Why is it possible to drop the permutation matrix? Answer the question and then confirm in Python using random matrices that $\mathbf{A}^\mathrm{T}\mathbf{A} = \mathbf{U}^\mathrm{T}\mathbf{L}^\mathrm{T}\mathbf{LU}$ even when $\mathbf{P} \neq \mathbf{I}$.

## CHAPTER 11

General Linear Models and Least Squares

The universe is a really big and really complicated place. All animals on Earth have a natural curiousity to explore and try to understand their environment, but we humans are privileged with the intelligence to develop scientific and statistical tools to take our curiosity to the next level. That's why we have airplanes, MRI machines, rovers on Mars, vaccines, and, of course, books like this one.

How do we understand the universe? By developing mathematically grounded theories, and by collecting data to test and improve those theories. And this brings us to statistical models. A statistical model is a simplified mathematical representation of some aspect of the world. Some statistical models are simple (e.g., predicting that the stock market will increase over decades); others are much more sophisticated, like the Blue Brain Project that simulates brain activity with such exquisite detail that one second of simulated activity requires 40 minutes of computation time.

A key distinction of *statistical* models (as opposed to other mathematical models) is that they contain free parameters that are fit to data. For example, I know that the stock market will go up over time, but I don't know by how much. Therefore, I allow the change in stock market price over time (that is, the slope) to be a free parameter whose numerical value is determined by data.

Crafting a statistical model can be difficult and requires creativity, experience, and expertise. But finding the free parameters based on fitting the model to data is a simple matter of linear algebra—in fact, you already know all the math you need for this chapter; it's just a matter of putting the pieces together and learning the statistics terminology.

General Linear Models

A statistical model is a set of equations that relates predictors (called *independent variables*) to observations (called the *dependent variable*). In the model of the stock market, the independent variable is *time* and the dependent variable is *stock market price* (e.g., quantified as the S&P 500 index).

In this book I will focus on general linear models, which are abbreviated as GLM. Regression is a type of GLM, for example.

Terminology

Statisticians use slightly different terminology than do linear algebraticians. Table 11-1 shows the key letters and descriptions for vectors and matrices used in the GLM.

Table 11-1. Table of terms in GLMs

LinAlg	Stats	Description
Ax = b	$X\beta = y$	General linear model (GLM)
A	X	Design matrix (columns = independent variables, predictors, regressors)
x	β	Regression coefficients or beta parameters
b	y	Dependent variable, outcome measure, data

Setting Up a General Linear Model

Setting up a GLM involves (1) defining an equation that relates the predictor variables to the dependent variable, (2) mapping the observed data onto the equations, (3) transforming the series of equations into a matrix equation, and (4) solving that equation.

I'll use a simple example to make the procedure concrete. I have a model that predicts adult height based on weight and on parents' height. The equation looks like this:

$$y = \beta_0 + \beta_1 w + \beta_2 h + \epsilon$$

y is the height of an individual, w is their weight, and h is their parents' height (the average of mother and father). ϵ is an error term (also called a *residual*), because we cannot reasonably expect that weight and parents' height *perfectly determines* an individual's height; there are myriad factors that our model does not account for, and the variance not attributable to weight and parents' height will be absorbed by the residual.

My hypothesis is that weight and parents' height are important for an individual's height, but I don't know *how* important each variable is. Enter the β terms: they are the coefficients, or weights, that tell me how to combine weight and parents' height to predict an individual's height. In other words, a linear weighted combination, where the βs are the weights.

β_0 is called an *intercept* (sometimes called a *constant*). The intercept term is a vector of all 1s. Without an intercept term, the best-fit line is forced to pass through the origin. I'll explain why, and show a demonstration, toward the end of the chapter.

Now we have our equation, our model of the universe (well, one tiny part of it). Next, we need to map the observed data onto the equations. For simplicity, I'm going to make up some data in Table 11-2 (you may imagine that y and h have units of centimeters and w has units of kilograms).

Table 11-2. Made-up data for our statistical model of height

y	w	h
175	70	177
181	86	190
159	63	180
165	62	172

Mapping the observed data onto our statistical model involves replicating the equation four times (corresponding to four observations in our dataset), each time replacing the variables y, w, and h with the measured data:

$$175 = \beta_0 + 70\beta_1 + 177\beta_2$$
$$181 = \beta_0 + 86\beta_1 + 190\beta_2$$
$$159 = \beta_0 + 63\beta_1 + 180\beta_2$$
$$165 = \beta_0 + 62\beta_1 + 172\beta_2$$

I'm omitting the ϵ term for now; I'll have more to say about the residuals later. We now need to translate this system of equations into a matrix equation. I know that you know how to do that, so I'll print out the equation here only so you can confirm what you already know from Chapter 10:

$$\begin{bmatrix} 1 & 70 & 177 \\ 1 & 86 & 190 \\ 1 & 63 & 180 \\ 1 & 62 & 172 \end{bmatrix} \begin{bmatrix} \beta_0 \\ \beta_1 \\ \beta_2 \end{bmatrix} = \begin{bmatrix} 175 \\ 181 \\ 159 \\ 165 \end{bmatrix}$$

And, of course, we can express this equation succinctly as $\mathbf{X}\boldsymbol{\beta} = \mathbf{y}$.

Solving GLMs

I'm sure you already know the main idea of this section: to solve for the vector of unknown coefficients β, simply left-multiply both sides of the equation by the left-inverse of $\mathbf{X}$, the design matrix. The solution looks like this:

$$\mathbf{X}\beta = \mathbf{y}$$
$$\left(\mathbf{X}^T\mathbf{X}\right)^{-1}\mathbf{X}^T\mathbf{X}\beta = \left(\mathbf{X}^T\mathbf{X}\right)^{-1}\mathbf{X}^T\mathbf{y}$$
$$\beta = \left(\mathbf{X}^T\mathbf{X}\right)^{-1}\mathbf{X}^T\mathbf{y}$$

Please stare at that final equation until it is permanently tattooed into your brain. It is called the *least squares solution* and is one of the most important mathematical equations in applied linear algebra. You'll see it in research publications, textbooks, blogs, lectures, docstrings in Python functions, billboards in Tajikistan,[1] and many other places. You might see different letters, or possibly some additions, like the following:

$$\mathbf{b} = \left(\mathbf{H}^T\mathbf{W}\mathbf{H} + \lambda\mathbf{L}^T\mathbf{L}\right)^{-1}\mathbf{H}^T\mathbf{x}$$

The meaning of that equation and the interpretation of the additional matrices are not important (they are various ways of regularizing the model fitting); what is important is that you are able to see the least squares formula embedded in that complicated-looking equation (for example, imagine setting $\mathbf{W} = \mathbf{I}$ and $\lambda = 0$).

The least squares solution via the left-inverse can be translated directly into Python code (variable X is the design matrix and variable y is the data vector):

```
X_leftinv = np.linalg.inv(X.T@X) @ X.T

# solve for the coefficients
beta = X_leftinv @ y
```

I will show results from these models—and how to interpret them—later in this chapter; for now I'd like you to focus on how the math formulas are translated into Python code.

---

1 Admittedly, I've never seen this equation on a Tajikistani billboard, but the point is to keep an open mind.

Left-Inverse Versus NumPy's Least Squares Solver

The code in this chapter is a direct translation of the math into Python code. Explicitly computing the left-inverse is not the most numerically stable way to solve the GLM (although it is accurate for the simple problems in this chapter), but I want you to see that the seemingly abstract linear algebra really works. There are more numerically stable ways to solve the GLM, including QR decomposition (which you will see later in this chapter) and Python's more numerically stable methods (which you will see in the next chapter).

Is the Solution Exact?

The equation $\mathbf{X}\boldsymbol{\beta} = \mathbf{y}$ is exactly solvable when $\mathbf{y}$ is in the column space of the design matrix $\mathbf{X}$. So the question is whether the data vector is guaranteed to be in the column space of the statistical model. The answer is no, there is no such guarantee. In fact, the data vector $\mathbf{y}$ is almost never in the column space of $\mathbf{X}$.

To understand why not, let's imagine a survey of university students in which the researchers are trying to predict average GPA (grade point average) based on drinking behavior. The survey may contain data from two thousand students yet have only three questions (e.g., how much alcohol do you consume; how often do you black out; what is your GPA). The data is contained in a 2000 × 3 table. The column space of the design matrix is a 2D subspace inside that 2000D ambient dimensionality, and the data vector is a 1D subspace inside that same ambient dimensionality.

If the data is in the column space of the design matrix, it means that the model accounts for 100% of the variance in the data. But this almost never happens: real-world data contains noise and sampling variability, and the models are simplifications that do not account for all of the variability (e.g., GPA is determined by myriad factors that our model ignores).

The solution to this conundrum is that we modify the GLM equation to allow for a discrepancy between the model-predicted data and the observed data. It can be expressed in several equivalent (up to a sign) ways:

$$\mathbf{X}\boldsymbol{\beta} = \mathbf{y} + \boldsymbol{\epsilon}$$
$$\mathbf{X}\boldsymbol{\beta} + \boldsymbol{\epsilon} = \mathbf{y}$$
$$\boldsymbol{\epsilon} = \mathbf{X}\boldsymbol{\beta} - \mathbf{y}$$

The interpretation of the first equation is that $\boldsymbol{\epsilon}$ is a residual, or an error term, that you add to the data vector so that it fits inside the column space of the design matrix. The interpretation of the second equation is that the residual term is an adjustment to the design matrix so that it fits the data perfectly. Finally, the interpretation of the third equation is that the residual is defined as the difference between the model-predicted data and the observed data.

There is another, very insightful, interpretation, which approaches GLMs and least squares from a geometric perspective. I'll get back to this in the next section.

The point of this section is that the observed data is almost never inside the subspace spanned by the regressors. For this reason, it is also common to see the GLM expressed as $\mathbf{X} = \beta\widehat{\mathbf{y}}$ where $\widehat{\mathbf{y}} = \mathbf{y} + \epsilon$.

Therefore, the goal of the GLM is to find the linear combination of the regressors that gets as close as possible to the observed data. More on this point later; I now want to introduce you to the geometric perspective of least squares.

A Geometric Perspective on Least Squares

So far I've introduced the solution to the GLM from the algebraic perspective of solving a matrix equation. There is also a geometric perspective to the GLM, which provides an alternative perspective and helps reveal several important features of the least squares solution.

Let's consider that the column space of the design matrix $C(\mathbf{X})$ is a subspace of $\mathbb{R}^M$. It's typically a very low-dimensional subspace (that is, $N << M$), because statistical models tend to have many more observations (rows) than predictors (columns). The dependent variable is a vector $\mathbf{y} \in \mathbb{R}^M$. The questions at hand are, is the vector $\mathbf{y}$ in the column space of the design matrix, and if not, what coordinate inside the column space of the design matrix is as close as possible to the data vector?

The answer to the first question is no, as I discussed in the previous section. The second question is profound, because you already learned the answer in Chapter 2. Consider Figure 11-1 while thinking about the solution.

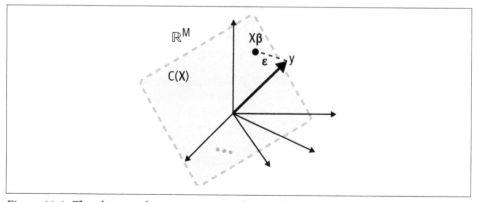

Figure 11-1. The abstracted geometric view of GLM: find the point in the column space of the design matrix that is as close as possible to the data vector

So, our goal is to find the set of coefficients β such that the weighted combination of columns in X minimizes the distance to data vector y. We can call that projection vector ϵ. How do we find the vector ϵ and the coefficients β? We use orthogonal vector projection, just like what you learned in Chapter 2. This means we can apply the same approach as in Chapter 2, but using matrices instead of vectors. The key insight is that the shortest distance between y and X is given by the projection vector $y - X\beta$ that meets X at a right angle:

$$X^T\epsilon = 0$$
$$X^T(y - X\beta) = 0$$
$$X^Ty - X^TX\beta = 0$$
$$X^TX\beta = X^Ty$$
$$\beta = \left(X^TX\right)^{-1}X^Ty$$

That progression of equations is remarkable: we started from thinking about the GLM as a geometric projection of a data vector onto the column space of a design matrix, applied the principle of orthogonal vector projection that you learned about early on in the book, and voilà! We have rederived the same left-inverse solution that we got from the algebraic approach.

Why Does Least Squares Work?

Why is it called "least squares"? What are these so-called squares, and why does this method give us the least of them?

The "squares" here refers to squared errors between the predicted data and the observed data. There is an error term for each ith predicted data point, which is defined as $\epsilon_i = X_i\beta - y_i$. Note that each data point is predicted using the same set of coefficients (that is, the same weights for combining the predictors in the design matrix). We can capture all errors in one vector:

$$\epsilon = X\beta - y$$

If the model is a good fit to the data, then the errors should be small. Therefore, we can say that the objective of model fitting is to choose the elements in β that minimize the elements in ϵ. But just *minimizing* the errors would cause the model to predict values toward negative infinity. Thus, instead we minimize the *squared* errors, which correspond to their geometric squared distance to the observed data y, regardless of whether the prediction error itself is positive or negative.[2] This is the same thing as minimizing the squared norm of the errors. Hence the name "least squares." That leads to the following modification:

$$\| \boldsymbol{\epsilon} \|^2 = \| \mathbf{X}\boldsymbol{\beta} - \mathbf{y} \|^2$$

We can now view this as an optimization problem. In particular, we want to find the set of coefficients that minimizes the squared errors. That minimization can be expressed as follows:

$$\min_{\boldsymbol{\beta}} \| \mathbf{X}\boldsymbol{\beta} - \mathbf{y} \|^2$$

The solution to this optimization can be found by setting the derivative of the objective to zero and applying a bit of differential calculus[3] and a bit of algebra:

$$0 = \frac{d}{d\boldsymbol{\beta}} \| \mathbf{X}\boldsymbol{\beta} - \mathbf{y} \|^2 = 2\mathbf{X}^{\mathrm{T}}(\mathbf{X}\boldsymbol{\beta} - \mathbf{y})$$

$$0 = \mathbf{X}^{\mathrm{T}}\mathbf{X}\boldsymbol{\beta} - \mathbf{X}^{\mathrm{T}}\mathbf{y}$$

$$\mathbf{X}^{\mathrm{T}}\mathbf{X}\boldsymbol{\beta} = \mathbf{X}^{\mathrm{T}}\mathbf{y}$$

$$\boldsymbol{\beta} = \left(\mathbf{X}^{\mathrm{T}}\mathbf{X}\right)^{-1}\mathbf{X}^{\mathrm{T}}\mathbf{y}$$

Amazingly enough, we started from a different perspective—minimize the squared distance between the model-predicted values and the observed values—and again we rediscovered the same solution to least squares that we reached simply by using our linear algebra intuition.

Figure 11-2 shows some observed data (black squares), their model-predicted values (gray dots), and the distances between them (gray dashed lines). All model-predicted values lie on a line; the goal of least squares is to find the slope and intercept of that line that minimizes the distances from predicted to observed data.

---

2 In case you were wondering, it is also possible to minimize the *absolute* distances instead of the *squared* distances. Those two objectives can lead to different results; one advantage of squared distance is the convenient derivative, which leads to the least squares solution.

3 If you are not comfortable with matrix calculus, then don't worry about the equations; the point is that we took the derivative with respect to $\boldsymbol{\beta}$ using the chain rule.

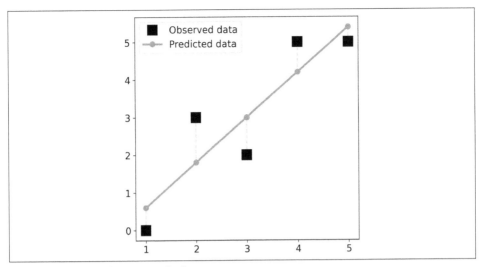

Figure 11-2. Visual intuition for least squares

All Roads Lead to Least Squares

You've now seen three ways to derive the least squares solution. Remarkably, all approaches lead to the same conclusion: left-multiply both sides of the GLM equation by the left-inverse of the design matrix **X**. Different approaches have unique theoretical perspectives that provide insight into the nature and optimality of least squares. But it is a beautiful thing that no matter how you begin your adventure into linear model fitting, you end up at the same conclusion.

GLM in a Simple Example

You will see several examples with real data in the next chapter; here I want to focus on a simple example with fake data. The fake data comes from a fake experiment in which I surveyed a random set of 20 of my fake students and asked them to report the number of my online courses they have taken and their general satisfaction with life.[4]

---

4 In case it's not already clear enough: this is completely made-up data for this example; any resemblance to the real world is coincidental.

Table 11-3 shows the first 4 (out of 20) rows of the data matrix.

Table 11-3. Data table

Number of courses	Life happiness
4	25
12	54
3	21
14	80

The data is easier to visualize in a scatterplot, which you see in Figure 11-3.

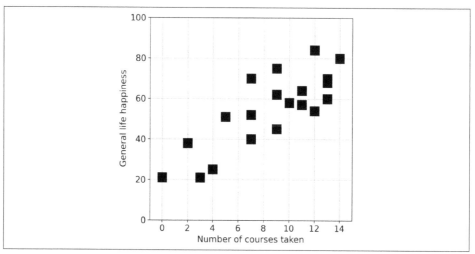

Figure 11-3. Fake data from a fake survey

Notice that the independent variable is plotted on the *x*-axis while the dependent variable is plotted on the *y*-axis. That is common convention in statistics.

We need to create the design matrix. Because this is a simple model with only one predictor, our design matrix is actually only one column vector. Our matrix equation $X\beta = y$ looks like this (again, only the first four data values):

$$\begin{bmatrix} 4 \\ 12 \\ 3 \\ 14 \end{bmatrix} [\beta] = \begin{bmatrix} 25 \\ 54 \\ 21 \\ 80 \end{bmatrix}$$

The following Python code shows the solution. Variables `numcourses` and `happiness` contain the data; they are both lists and therefore must be converted into multidimensional NumPy arrays:

```
X = np.array(numcourses,ndmin=2).T

# compute the left-inverse
X_leftinv = np.linalg.inv(X.T@X) @ X.T

# solve for the coefficients
beta = X_leftinv @ happiness
```

The least squares formula tells us that β = 5.95. What does this number mean? It is the slope in our formula. In other words, for each additional course that someone takes, their self-reported life happiness increases by 5.95 points. Let's see how that result looks in a plot. Figure 11-4 shows the data (black squares), the predicted happiness values (gray dots connected by a line), and the residuals (dashed line connecting each observed value to the predicted value).

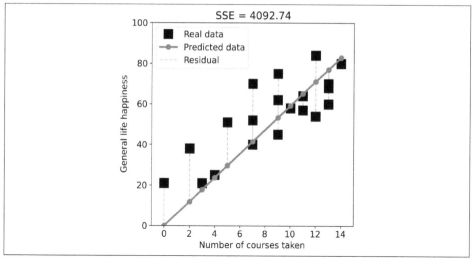

Figure 11-4. Observed and predicted data (SSE = sum of squared errors)

If you experience a feeling of unease while looking at Figure 11-4, then that's good—it means you are thinking critically and noticed that the model doesn't do a great job at minimizing the errors. You can easily imagine pushing the left side of the best-fit line up to get a better fit. What's the problem here?

The problem is that the design matrix contains no intercept. The equation of the best-fit line is $y = mx$, which means that when $x = 0$, $y = 0$. That constraint doesn't make sense for this problem—it would be a sad world if anyone who doesn't take my courses is completely devoid of life satisfaction. Instead, we want our line to have the form $y = mx + b$, where b is the intercept term that allows the best-fit line to cross the y-axis at any value. The statistical interpretation of the intercept is the expected numerical value of the observations when the predictors are set to zero.

Adding an intercept term to our design matrix gives the following modified equations (again only showing the first four rows):

$$\begin{bmatrix} 1 & 4 \\ 1 & 12 \\ 1 & 3 \\ 1 & 14 \end{bmatrix} \begin{bmatrix} \beta_1 \\ \beta_2 \end{bmatrix} = \begin{bmatrix} 25 \\ 54 \\ 21 \\ 80 \end{bmatrix}$$

The code doesn't change, with one exception of creating the design matrix:

```
X = np.hstack((np.ones((20,1)),np.array(numcourses,ndmin=2).T))
```

Now we find that β is the two-element vector [22.9,3.7]. The expected level of happiness for someone who has taken zero courses is 22.9, and for each additional course someone takes, their happiness increases by 3.7 points. I'm sure you will agree that Figure 11-5 looks much better. And the SSE is around half of what it was when we excluded the intercept.

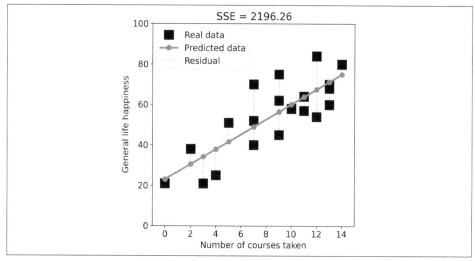

Figure 11-5. Observed and predicted data, now with an intercept term

I will let you draw your own conclusions about the fake results from this fake study based on fake data; the point is to see a numerical example of how to solve a system of equations by building an appropriate design matrix and solving for the unknown regressors using the left-inverse.

Least Squares via QR

The left-inverse method is theoretically sound, but risks numerical instability. This is partly because it requires computing the matrix inverse, which you already know can be numerically unstable. But it turns out that the matrix X^TX itself can introduce difficulties. Multiplying a matrix by its transpose has implications for properties such as the norm and the condition number. You will learn more about the condition number in Chapter 14, but I've already mentioned that matrices with a high condition number can be numerically unstable, and thus a design matrix with a high condition number will become even less numerically stable when squared.

QR decomposition provides a more stable way to solve the least squares problem. Observe the following sequence of equations:

$$X\beta = y$$
$$QR\beta = y$$
$$R\beta = Q^Ty$$
$$\beta = R^{-1}Q^Ty$$

These equations are slightly simplified from the actual low-level numerical implementations. For example, R is the same shape as X, i.e., tall (and therefore noninvertible), although only the first N rows are nonzero (as you discovered in Exercise 9-7), which means that rows $N + 1$ through M do not contribute to the solution (in matrix multiplication, rows of zeros produce results of zeros). Those rows can be removed from R and from Q^Ty. Secondly, row swaps (implemented via permutation matrices) might be used to increase numerical stability.

But here's the best part: it's not even necessary to invert R—that matrix is upper-triangular and therefore the solution can be obtained via back substitution. It's the same as solving a system of equations via the Gauss-Jordan method: augment the coefficients matrix by the constants, row reduce to obtain the RREF, and extract the solution from the final column of the augmented matrix.

The conclusion here is that QR decomposition solves the least squares problem without squaring X^TX and without explicitly inverting a matrix. The main risk of numerical instability comes from computing Q, although this is fairly numerically stable when implemented via Householder reflections.

Exercise 11-3 will walk you through this implementation.

Summary

Many people think that statistics is hard because the underlying mathematics is hard. To be sure, there are advanced statistical methods involving advanced mathematics. But many of the commonly used statistical methods are built on linear algebra principles that you now understand. That means you no longer have any excuses not to master the statistical analyses used in data science!

The goal of this chapter was to introduce you to the terminology and math underlying the general linear model, the geometric interpretation, and the implications of the math for minimizing the difference between model-predicated data and observed data. I also showed an application of a regression in a simple toy example. In the next chapter, you'll see least squares implemented with real data, and you'll see extensions of least squares in regression, such as polynomial regression and regularization.

Here are the key take-home messages from this chapter:

- The GLM is a statistical framework for understanding our rich and beautiful universe. It works by setting up a system of equations, just like the systems of equations you learned about in the previous chapter.

- The terms are somewhat different between linear algebra and statistics; once you learn the terminological mappings, statistics becomes easier because you already know the math.

- The least squares method of solving equations via the left-inverse is the foundation of many statistical analyses, and you will often see the least squares solution "hidden" inside seemingly complicated formulas.

- The least squares formula can be derived via algebra, geometry, or calculus. This provides multiple ways of understanding and interpreting least squares.

- Multiplying the observed data vector by the left-inverse is conceptually the right way to think about least squares. In practice, other methods such as LU and QR decomposition are more numerically stable. Fortunately, you don't need to worry about this because Python calls low-level libraries (mostly LAPACK) that implement the most numerically stable algorithms.

Code Exercises

Exercise 11-1.

I explained that the residuals are orthogonal to the predicted data (in other words, $\epsilon^T \hat{y} = 0$). Illustrate this in the toy dataset from this chapter. In particular, make a scatterplot of the predicted data by the errors (as in Figure 11-6). Then compute the dot product and the correlation coefficient between the residuals and the model-predicted data. In theory, both should be exactly zero, although there are some rounding errors. Which of those two analyses (dot product or correlation) is smaller, and why is that?

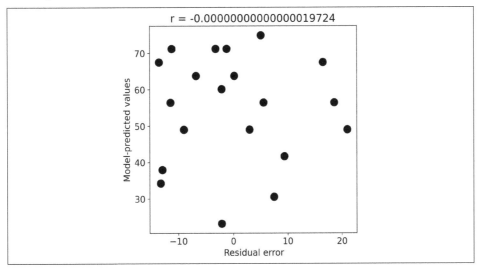

Figure 11-6. Solution to Exercise 11-1

Exercise 11-2.

The model-predicted happiness is merely one way of linearly combining the columns of the design matrix. But the residuals vector isn't only orthogonal to that one linear weighted combination; instead, the residuals vector is orthogonal to the *entire subspace* spanned by the design matrix. Demonstrate this in Python (hint: think of the left-null space and rank).

Exercise 11-3.

You are now going to compute least squares via QR decomposition, as I explained in "Least Squares via QR" on page 187. In particular, compute and compare the following solution methods: (1) the left-inverse $\left(\mathbf{X}^T\mathbf{X}\right)^{-1}\mathbf{X}^T\mathbf{y}$, (2) QR with the inverse as $\mathbf{R}^{-1}\mathbf{Q}^T\mathbf{y}$, and (3) Gauss-Jordan elimination on the matrix $\mathbf{R}$ augmented with $\mathbf{Q}^T\mathbf{y}$.

Print out the beta parameters from the three methods as follows. (Rounding to three digits after the decimal point is an optional extra coding challenge.)

```
Betas from left-inverse:
[23.13   3.698]

Betas from QR with inv(R):
[23.13   3.698]
```

```
Betas from QR with back substitution:
[[23.13    3.698]]
```

Finally, print out the resulting matrices from the QR method as follows:

```
Matrix R:
[[ -4.472 -38.237]
 [  0.      17.747]]

Matrix R|Q'y:
[[  -4.472  -38.237 -244.849]
 [   0.       17.747   65.631]]

Matrix RREF(R|Q'y):
[[ 1.    0.     23.13 ]
 [ 0.    1.      3.698]]
```

Exercise 11-4.

Outliers are data values that are unusual or nonrepresentative. Outliers can cause significant problems in statistical models, and therefore can cause significant headaches for data scientists. In this exercise, we will create outliers in the happiness data to observe the effects on the resulting least squares solution.

In the data vector, change the first observed data point from 70 to 170 (simulating a data-entry typo). Then recompute the least squares fit and plot the data. Repeat this outlier simulation but change the final data point from 70 to 170 (and set the first data point back to its original 70). Compare with the original data by creating a visualization like Figure 11-7.

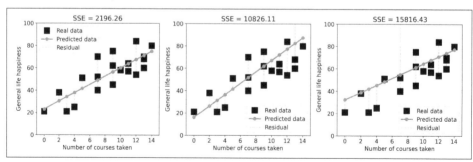

Figure 11-7. Solution to Exercise 11-4

Interestingly, the outlier was identical in the outcome variable (in both cases, the 70 turned into 170), but the effect on the fit of the model to the data was quite different because of the corresponding x-axis value. This differential impact of outliers is called *leverage* and is something you would learn about in a deeper discussion of statistics and model fitting.

Exercise 11-5.

In this exercise, you will compute the matrix inverse using least squares, following the interpetation I introduced in the previous chapter. We will consider the equation $\mathbf{XB} = \mathbf{Y}$, where $\mathbf{X}$ is the square full-rank matrix to invert, $\mathbf{B}$ is the matrix of unknown coefficients (which will be the matrix inverse), and $\mathbf{Y}$ is the "observed data" (the identity matrix).

You will compute $\mathbf{B}$ in three ways. First, use the left-inverse least squares method to compute the matrix one column at a time. This is done by computing the least squares fit between the matrix $\mathbf{X}$ and each column of $\mathbf{Y}$ in a for loop. Second, use the left-inverse method to compute the entire $\mathbf{B}$ matrix in one line of code. Finally, compute $\mathbf{X}^{-1}$ using the function np.linalg.inv(). Mutliply each of those $\mathbf{B}$ matrices by $\mathbf{X}$ and show in a figure like Figure 11-8. Finally, test whether these three "different" ways of computing the inverse are equivalent (they should be, because the matrix inverse is unique).

Observation: it is rather strange (not to mention circular) to use the inverse of $\mathbf{X}^T\mathbf{X}$ to compute the inverse of $\mathbf{X}$. (Indeed, you should confirm with paper and pencil that the left-inverse reduces to the full inverse for square full-rank $\mathbf{X}$!) Needless to say, this is not a computational method that would be implemented in practice. However, this exercise reinforces the interpretation of the matrix inverse as the transformation that projects a matrix onto the identity matrix, and that that projection matrix can be obtained via least squares. Comparing the least squares solution to np.linalg.inv also illustrates the numerical inaccuracies that can arise when computing the left-inverse.

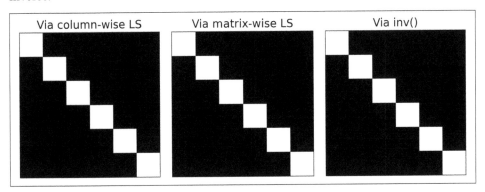

Figure 11-8. Solution to Exercise 11-5

## CHAPTER 12

Least Squares Applications

In this chapter, you will see a few applications of least squares model fitting in real data. Along the way, you will learn how to implement least squares using several different—and more numerically stable—Python functions, and you will learn some new concepts in statistics and machine learning such as multicollinearity, polynomial regression, and the grid search algorithm as an alternative to least squares.

By the end of this chapter, you will have a deeper understanding of how least squares is used in applications, including the importance of numerically stable algorithms for "difficult" situations involving reduced-rank design matrices. And you will see that the analytic solution provided by least squares outperforms an empirical parameter search method.

Predicting Bike Rentals Based on Weather

I'm a big fan of bicycles and a big fan of bibimbap (a Korean dish made with rice and veggies or meat). Therefore, I was happy to find a publicly available dataset about bike rentals in Seoul.[1] The dataset contains nearly nine thousand observations of data about the number of bikes that were rented in the city and variables about the weather including temperature, humidity, rainfall, windspeed, and so on.

The purpose of the dataset is to predict the demand for bike sharing based on weather and season. That is important because it will help bike rental companies and local governments optimize the availability of healthier modes of transportation. It's a great dataset, there's a lot that could be done with it, and I encourage you to spend

---

1 V E Sathishkumar, Jangwoo Park, and Yongyun Cho, "Using Data Mining Techniques for Bike Sharing Demand Prediction in Metropolitan City," *Computer Communications*, 153, (March 2020): 353–366, data downloaded from *https://archive.ics.uci.edu/ml/datasets/Seoul+Bike+Sharing+Demand*.

time exploring it. In this chapter, I will focus on building relatively simple regression models to predict bike rental counts based on a few features.

Although this is a book on linear algebra and not statistics, it is still important to inspect the data carefully before applying and interpreting statistical analyses. The online code has more details about importing and inspecting the data using the pandas library. Figure 12-1 shows the data from bike count rentals (the dependent variable) and rainfall (one of the independent variables).

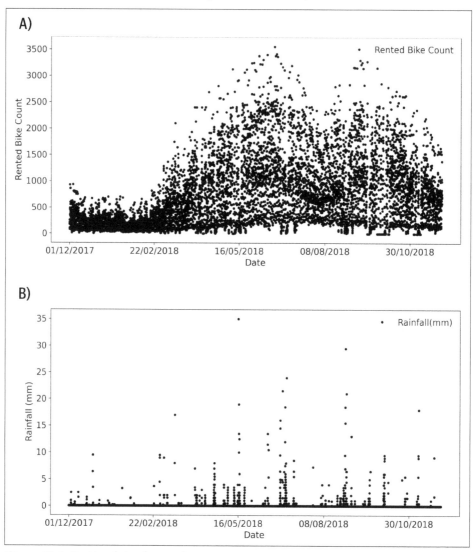

Figure 12-1. Scatterplots of some data

Notice that rainfall is a sparse variable—it's mostly zeros with a relatively small number of nonzero values. We'll come back to this in the exercises.

Figure 12-2 shows a correlation matrix from four selected variables. Inspecting correlation matrices is always a good idea before starting statistical analyses, because it will show which variables (if any) are correlated and can reveal errors in the data (e.g., if two supposedly different variables are perfectly correlated). In this case, we see that bike rental count is positively correlated with *hour* and *temperature* (people rent more bikes later in the day and when the weather is warmer) and negatively correlated with *rainfall*. (Note that I'm not showing statistical significance here, so these interpretations are qualitative.)

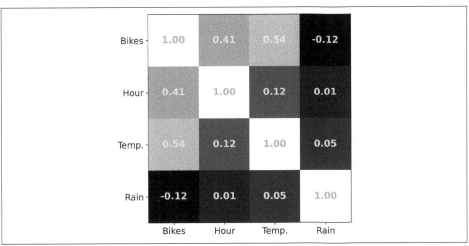

Figure 12-2. Correlation matrix of four selected variables

In the first analysis, I want to predict bike rental counts based on rainfall and the seasons. The seasons (winter, spring, summer, fall) are text labels in the dataset, and we need to convert them to numbers for the analysis. We could translate the four seasons into the numbers 1–4, but seasons are circular while regressions are linear. There are a few ways to deal with this, including using an ANOVA instead of a regression, using one-hot-encoding (used in deep learning models), or binarizing the seasons. I'm going to take the latter approach and label autumn and winter "0" and spring and summer "1". The interpretation is that a positive beta coefficient indicates more bike rentals in spring/summer compared to autumn/winter.

(Tangential note: on the one hand, I could have made things simpler by selecting only continuous variables. But I want to stress that there is more to data science than just applying a formula to a dataset; there are many nontrivial decisions that affect the kinds of analyses you can do, and therefore the kinds of results you can obtain.)

The left side of Figure 12-3 shows the design matrix visualized as an image. This is a common representation of the design matrix, so make sure you are comfortable interpreting it. The columns are regressors and the rows are observations. Columns are sometimes normalized to facilitate visual interpretation if the regressors are in very different numerical scales, although I didn't do that here. You can see that rainfall is sparse and that the dataset spans two autumn/winter periods (black areas in the middle column) and one spring/summer period (white area in the middle). The intercept is, of course, solid white, because it takes the same value for every observation.

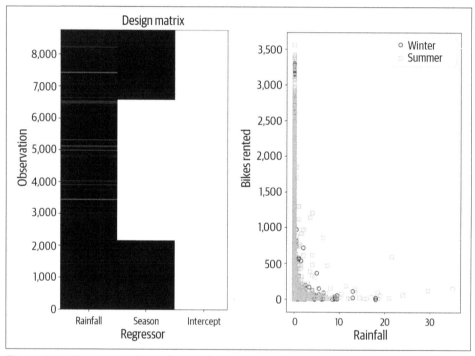

Figure 12-3. Design matrix and some data

The right side of Figure 12-3 shows the data, plotted as rainfall by bikes rented separately for the two seasons. Clearly, the data do not lie on a line, because there are many values at or close to zero on both axes. In other words, visually inspecting the data suggests that the relationships amongst the variables are nonlinear, which means that a linear modeling approach might be suboptimal. Again, this highlights the importance of visually inspecting data and carefully selecting an appropriate statistical model.

Nonetheless, we will forge ahead using a linear model fit to the data using least squares. The following code shows how I created the design matrix (variable `data` is a pandas dataframe):

```
# Create a design matrix and add an intercept
desmat = data[['Rainfall(mm)','Seasons']].to_numpy()
desmat = np.append(desmat,np.ones((desmat.shape[0],1)),axis=1)

# extract DV
y = data[['Rented Bike Count']].to_numpy()

# fit the model to the data using least squares
beta = np.linalg.lstsq(desmat,y,rcond=None)
```

The beta values for *rainfall* and *season* are, respectively, −80 and 369. These numbers indicate that there are fewer bike rentals when it rains and that there are more bike rentals in the spring/summer compared to autumn/winter.

Figure 12-4 shows the predicted versus the observed data, separated for the two seasons. If the model were a perfect fit to the data, the dots would lie on a diagonal line with a slope of 1. Clearly, that's not the case, meaning that the model did not fit the data very well. Indeed, the R^2 is a paltry 0.097 (in other words, the statistical model accounts for around 1% of the variance in the data). Futhermore, you can see that the model predicts *negative* bike rentals, which is not interpretable—bike rental counts are strictly nonnegative numbers.

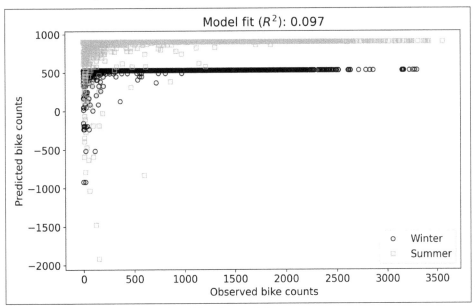

Figure 12-4. Scatterplot of predicted by observed data

Thus far in the code, we have received no warnings or errors; we have done nothing wrong in terms of math or coding. However, the statistical model we used is not the most appropriate for this research question. You will have the opportunity to improve it in Exercise 12-1 and Exercise 12-2.

Regression Table Using statsmodels

Without getting too deep into the statistics, I want to show you how to create a regression table using the statsmodels library. This library works with pandas dataframes instead of NumPy arrays. The following code shows how to set up and compute the regression model (OLS stands for *ordinary least squares*):

```
import statsmodels.api as sm

# extract data (staying with pandas dataframes)
desmat_df  = data[['Rainfall(mm)','Seasons']]
obsdata_df = data['Rented Bike Count']

# create and fit the model (must explicitly add intercept)
desmat_df = sm.add_constant(desmat_df)
model = sm.OLS(obsdata_df,desmat_df).fit()
print( model.summary() )
```

The regression table contains a lot of information. It's OK if you don't understand all of it; the key items you can look for are the R^2 and the regression coefficients (coef) for the regressors:

```
==============================================================================
Dep. Variable:     Rented Bike Count  R-squared:                       0.097
Model:                          OLS   Adj. R-squared:                  0.097
Method:               Least Squares   F-statistic:                     468.8
Date:              Wed, 26 Jan 2022   Prob (F-statistic):          3.80e-194
Time:                      08:40:31   Log-Likelihood:                -68654.
No. Observations:              8760   AIC:                         1.373e+05
Df Residuals:                  8757   BIC:                         1.373e+05
Df Model:                         2
Covariance Type:          nonrobust
==============================================================================
                 coef     std err          t      P>|t|      [0.025      0.975]
------------------------------------------------------------------------------
const          530.4946     9.313     56.963      0.000     512.239     548.750
Rainfall(mm)   -80.5237     5.818    -13.841      0.000     -91.928     -69.120
Seasons        369.1267    13.127     28.121      0.000     343.395     394.858
==============================================================================
Omnibus:                   1497.901   Durbin-Watson:                   0.240
Prob(Omnibus):                0.000   Jarque-Bera (JB):             2435.082
Skew:                         1.168   Prob(JB):                         0.00
Kurtosis:                     4.104   Cond. No.                         2.80
==============================================================================
```

Multicollinearity

If you've taken a statistics course, you might have heard of the term *multicollinearity*. The Wikipedia definition is "one predictor variable in a multiple regression model can be linearly predicted from the others with a substantial degree of accuracy."[2]

This means that there are linear dependencies in the design matrix. In the parlance of linear algebra, multicollinearity is just a fancy term for *linear dependence*, which is the same thing as saying that the design matrix is reduced-rank or that it is singular.

A reduced-rank design matrix does not have a left-inverse, which means that the least squares problem cannot be solved analytically. You will see the implications of multicollinearity in Exercise 12-3.

Solving GLMs with Multicollinearity

It is, in fact, possible to derive solutions for GLMs that have a reduced-rank design matrix. This is done using a modification to the QR procedure in the previous chapter and using the MP pseudoinverse. In the case of reduced-rank design matrix, there is no unique solution, but we can select the solution with the minimum error. This is called the *minimum norm solution*, or simply min-norm, and is used often, for example in biomedical imaging. Special applications notwithstanding, a design matrix with linear dependencies usually indicates a problem with the statistical model and should be investigated (data entry mistakes and coding bugs are common sources of multicollinearity).

Regularization

Regularization is an umbrella term that refers to various ways of modifying a statistical model, with the goal of improving numerical stability, transforming singular or ill-conditioned matrices to full-rank (and thus invertible), or improving generalizability by reducing overfitting. There are several forms of regularization depending on the nature of the problem and the goal of regularizing; some specific techniques you might have heard of include Ridge (a.k.a. L2), Lasso (a.k.a. L1), Tikhonov, and shrinkage.

Different regularization techniques work in different ways, but many regularizers "shift" the design matrix by some amount. You will recall from Chapter 5 that shifting a matrix means adding some constant to the diagonal as $\mathbf{A} + \lambda\mathbf{I}$, and from Chapter 6 that shifting a matrix can transform a reduced-rank matrix into a full-rank matrix.

---

2 Wikipedia, s.v. "multicollinearity," *https://en.wikipedia.org/wiki/Multicollinearity*.

In this chapter, we will regularize the design matrix by shifting it according to some proportion of its Frobenius norm. This modifies the least squares solution Equation 12-1.

Equation 12-1. Regularization

$$\beta = (\mathbf{X}^\mathrm{T}\mathbf{X} + \gamma \parallel \mathbf{X} \parallel_F^2 \mathbf{I})^{-1}\mathbf{X}^\mathrm{T}\mathbf{y}$$

The key parameter is γ (Greek letter *gamma*), which determines the amount of regularization (observe that $\gamma = 0$ corresponds to no regularization). Choosing an appropriate γ parameter is nontrivial, and is often done through statistical techniques like cross-validation.

The most obvious effect of regularization is that if the design matrix is reduced-rank, then the regularized squared design matrix is full-rank. Regularization also decreases the condition number, which measures the "spread" of information in the matrix (it's the ratio of the largest to the smallest singular value; you'll learn about this in Chapter 14). This increases the numerical stability of the matrix. The statistical implication of regularization is to "smooth out" the solution by reducing the sensitivity of the model to individual data points that might be outliers or nonrepresentative, and therefore less likely to be observed in new datasets.

Why do I scale by the squared Frobenius norm? Consider that a specified value of γ, say, $\gamma = .01$, can have a huge or a negligible impact on the design matrix depending on the range of numerical values in the matrix. Therefore, we scale to the numerical range of the matrix, which means we interpret the γ parameter as the *proportion* of regularization. The reason for squaring the Frobenius norm is that $\parallel \mathbf{X} \parallel_F^2 = \parallel \mathbf{X}^\mathrm{T}\mathbf{X} \parallel_F$. In other words, the squared norm of the design matrix equals the norm of the design matrix times its transpose.

It's actually more common to use the average of the eigenvalues of the design matrix instead of the Frobenius norm. After learning about eigenvalues in Chapter 13, you'll be able to compare the two regularization methods.

Implementing regularization in code is the focus of Exercise 12-4.

Polynomial Regression

A *polynomial regression* is like a normal regression but the independent variables are the x-axis values raised to higher powers. That is, each column i of the design matrix is defined as x^i, where x is typically time or space but can be other variables such as medication dosage or population. The mathematical model looks like this:

$$y = \beta_0 x^0 + \beta_1 x^1 + \ldots + \beta_n x^n$$

Note that $x^0 = 1$, giving us the intercept of the model. Otherwise, it's still a regular regression—the goal is to find the β values that minimize the squared differences between the predicted and observed data.

The *order* of the polynomial is the largest power i. For example, a fourth order polynomial regression has terms up to x^4 (if there is no x^3 term, then it's still a fourth order model with $\beta_3 = 0$).

Figure 12-5 shows an example of the individual regressors and the design matrix of a third order polynomial (keep in mind that an nth order polynomial has $n + 1$ regressors including the intercept). The polynomial functions are the basis vectors for modeling the observed data.

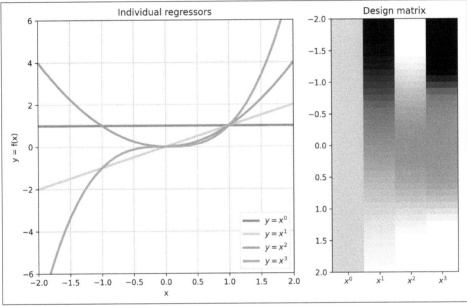

Figure 12-5. Design matrix of a polynomial regression

Other than the special design matrix, a polynomial regression is exactly the same as any other regression: use the left-inverse (or more computationally stable alternatives) to obtain the set of coefficients such that the weighted combination of regressors (i.e., the predicted data) best matches the observed data.

Polynomial regressions are used in curve fitting and in approximating nonlinear functions. Applications include time series modeling, population dynamics, dose-response functions in medical research, and physical stresses on structural support beams. Polynomials can also be expressed in 2D, which are used to model spatial structure such as earthquake propagation and brain activity.

Enough background. Let's work through an example. The dataset I picked is from a model of human population doubling. The question is "How long does it take for the population of humanity to double (e.g., from five hundred million to one billion)?" If the rate of population increase is itself increasing (because more people have more babies, who grow up to have even more babies), then the doubling time will decrease with each doubling. On the other hand, if the population growth slows (people have fewer babies), then the doubling time will increase over successive doublings.

I found a relevant dataset online.[3] It's a small dataset, so all the numbers are available in the online code and shown in Figure 12-6. This dataset includes both actual measured data and projections to the year 2100. These projections into the future are based on a number of assumptions, and no one really knows how the future will play out (that's why you should find a balance between preparing for the future and enjoying the moment). Regardless, the data thus far shows that the human population doubled with increasing frequency over the past five hundred years (at least), and the authors of the dataset predict that the doubling rate will increase slightly over the next century.

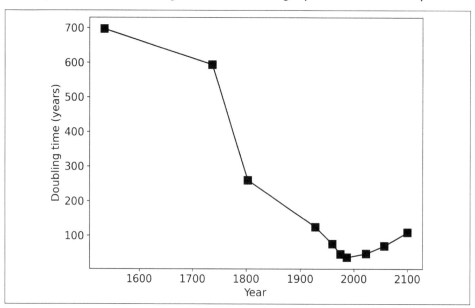

Figure 12-6. Plot of the data

I chose a third order polynomial to fit the data, and created and fit the model using the following code (variable `year` contains the *x*-axis coordinates and variable `doubleTime` contains the dependent variable):

3 Max Roser, Hannah Ritchie, and Esteban Ortiz-Ospina, "World Population Growth," OurWorldInData.org, 2013, *https://ourworldindata.org/world-population-growth*.

```
# design matrix
X = np.zeros((N,4))
for i in range(4):
  X[:,i] = np.array(year)**i

# fit the model and compute the predicted data
beta = np.linalg.lstsq(X,doubleTime, rcond=None)
yHat = X@beta[0]
```

Figure 12-7 shows the predicted data using the polynomial regression created by that code.

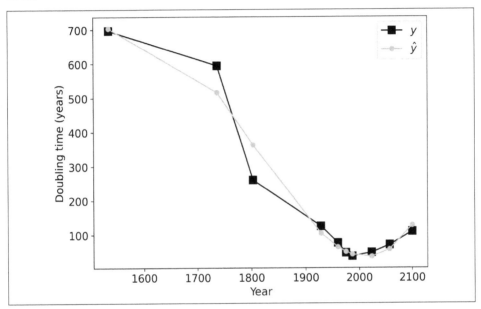

Figure 12-7. Plot of the data

The model captures both the downward trend and the projected upswing in the data. Without further statistical analysis, we cannot say that this is the *best* model or that the model is a statistically significantly good fit to the data. But it is clear that polynomial regressions are well suited for fitting curves. You will continue exploring this model and the data in Exercise 12-5, but I encourage you to play around with the code that produces Figure 12-7 by testing different order parameters.

Polynomial regressions are commonly used, and NumPy has dedicated functions to create and fit such models:

```
beta = np.polyfit(year,doubleTime,3) # 3rd order
yHat = np.polyval(beta,year)
```

Grid Search to Find Model Parameters

Least squares via the left-inverse is a brilliant way to fit models to data. Least squares is accurate, fast, and deterministic (meaning that each time you rerun the code, you'll get the same result). But it works only for linear model fitting, and not all models can be fit using linear methods.

In this section, I will introduce you to another optimization method used to identify model parameters, known as *grid search*. A grid search works by sampling the parameter space, computing the model fit to the data with each parameter value, and then selecting the parameter value that gave the best model fit.

As a simple example, let's take the function $y = x^2$. We want to find the minimum of that function. Of course, we already know that the minimum is at $x = 0$; this helps us understand and evaluate the results of the grid search method.

In the grid search technique, we start with a predefined set of x values to test. Let's use the set $(-2, -1, 0, 1, 2)$. That's our "grid." Then we compute the function at each of those grid values to obtain $y = (4, 1, 0, 1, 4)$. And we find that the minimum y occurs when $x = 0$. In this case, the grid-based solution is the same as the true solution.

But grid search is not guaranteed to give the optimal solution. For example, imagine our grid was $(-2, -.5, 1, 2.5)$; the function values would be $y = (4, .25, 1, 6.25)$, and we would conclude that $x = -.5$ is the parameter value that minimizes the function $y = x^2$. That conclusion is "kind of correct" because it is the best solution within the specified grid. Grid-search failures can also arise from a poorly chosen range of values. Imagine, for example, that our grid was $(-1000, -990, -980, -970)$; we would conclude that $y = x^2$ is minimized when $x = -970$.

The point is that both the range and the resolution (the spacing between grid points) are important, because they determine whether you will obtain the *best* solution, a *pretty good* solution, or an *awful* solution. In the toy example above, the appropriate range and resolution are easy to detemine. In complicated, multivariate, nonlinear models, appropriate grid search parameters might take some more work and exploration.

I ran a grid search on the "happy student" data from the previous chapter (as a reminder: it was fake data from a fake survey showing that people who enrolled in more of my courses had higher life satisfaction). The model of those data has two parameters—intercept and slope—and so we evaluate that function at each point on a 2D grid of possible parameter pairings. Results are shown in Figure 12-8.

What does that graph mean, and how do we interpret it? The two axes correspond to values of parameters, and therefore each coordinate in that graph creates a model with those corresponding parameter values. Then, the fit of each of those models to the data is calculated and stored, and visualized as an image.

The coordinate with the best fit to the data (the smallest sum of squared errors) is the optimal parameter set. Figure 12-8 also shows the analytic solution using the least squares approach. They are close but not exactly overlapping. In Exercise 12-6, you will have the opportunity to implement this grid search as well as to explore the importance of the grid resolution on the accuracy of the results.

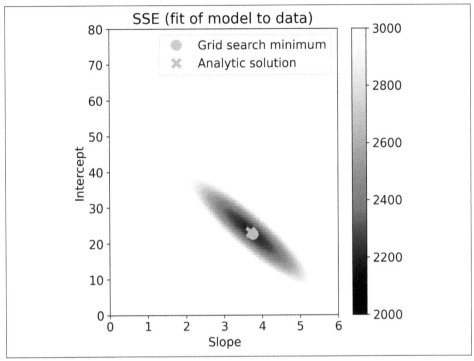

Figure 12-8. Results from a grid search on the "happy student" dataset. The intensity shows the sum of squared errors fit to the data.

Why would anyone use grid search when least squares is better and faster? Well, you should never use a grid search method when least squares is a viable solution. Grid search is a useful technique for finding parameters in nonlinear models and is often used, for example, to identify hyperparameters in deep learning models (*hyperparameters* are model architecture design features that are selected by the researcher, not learned from data). Grid search can be time-consuming for large models, but parallelization can make grid search more feasible.

The conclusion is that grid search is a nonlinear method for fitting models to data when linear methods cannot be applied. Along your journey into data science and machine learning, you will also learn about additional nonlinear methods, including simplex and the famous gradient descent algorithm that powers deep learning.

Summary

I hope you enjoyed reading about applications of least squares and the comparison to other model-fitting approaches. The following exercises are the most important part of this chapter, so I don't want to take up your time with a long summary. Here are the important points:

- Visual inspection of data is important for selecting the right statistical models and interpreting statistical results correctly.

- Linear algebra is used for quantitative assessments of datasets, including correlation matrices.

- The matrix visualization methods you learned in Chapter 5 are useful for inspecting design matrices.

- Mathematical concepts are sometimes given different names in different fields. An example in this chapter is *multicollinearity*, which means linear dependences in the design matrix.

- Regularization involves "shifting" the design matrix by some small amount, which can increase numerical stability and likelihood of generalizing the new data.

- Having a deep understanding of linear algebra can help you select the most appropriate statistical analyses, interpret the results, and anticipate potential problems.

- A polynomial regression is the same as a "regular" regression, but the columns in the design matrix are defined as the *x*-axis values raised to increasing powers. Polynomial regressions are used for curve fitting.

- Grid search is a nonlinear method of model fitting. Linear least squares is the optimal approach when the model is linear.

Code Exercises

Bike Rental Exercises

Exercise 12-1.

Perhaps part of the problem with negative bike rentals in Figure 12-4 can be alleviated by eliminating the no-rainfall days. Repeat the analysis and graph for this analysis, but select only the data rows that have nonzero rainfall. Do the results improve, in terms of higher R^2 and positive predicted rental counts?

Exercise 12-2.

Because *seasons* is a categorical variable, an ANOVA would actully be a more appropriate statistical model than a regression. Maybe the binarized *seasons* lacks the sensitivity to predict bike rentals (for example, there can be warm sunny days in autumn and cold rainy days in spring), and therefore temperature might be a better predictor.

Replace *seasons* with *temperature* in the design matrix and rerun the regression (you can use all days, not only the no-rainfall days from the previous exercise), and reproduce Figure 12-9. There is still the issue with predicting negative rentals (this is because of the linearity of the model), but the R^2 is higher and the prediction appears qualitatively better.

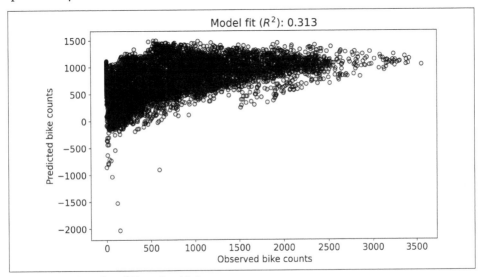

Figure 12-9. Results of Exercise 12-2

Multicollinearity Exercise

Exercise 12-3.

This exercise continues with the model from Exercise 12-2.[4] This model contains three regressors, including the intercept. Create a new design matrix that contains a fourth regressor defined as some linear weighted combination of *temperature* and *rainfall*. Give this design matrix a different variable name, because you will need it in

---

4 If you encounter Python errors, you might need to rerun previous code then re-create the design matrix variables.

the next exercise. Confirm that the design matrix has four columns yet a rank of 3, and compute the correlation matrix of the design matrix.

Note that depending on the weightings of the two variables, you wouldn't expect a correlation of 1 even with linear dependencies; you also wouldn't expect to reproduce the exact correlations here:

```
Design matrix size: (8760, 4)
Design matrix rank: 3

Design matrix correlation matrix:
[[1.        0.05028    nan 0.7057 ]
 [0.05028 1.          nan 0.74309]
 [   nan      nan      nan     nan]
 [0.7057  0.74309     nan 1.      ]]
```

Fit the model using three different coding approaches: (1) the direct implementation with the left-inverse as you learned in the previous chapter, (2) using NumPy's lstsq function, and (3) using statsmodels. For all three methods, compute R^2 and the regression coefficients. Print the results as follows. The numerical instability of np.linalg.inv on a reduced-rank design matrix is apparent.

```
MODEL FIT TO DATA:
  Left-inverse: 0.0615
  np lstsqr    : 0.3126
  statsmodels : 0.3126

BETA COEFFICIENTS:
  Left-inverse: [[-1528.071    11.277    337.483    5.537 ]]
  np lstsqr    : [[   -87.632     7.506    337.483    5.234 ]]
  statsmodels : [    -87.632     7.506    337.483    5.234 ]
```

Side note: no errors or warning messages were given; Python simply gave outputs even though there is clearly something wrong with the design matrix. We can debate the merits of this, but this example highlights—yet again—that understanding the linear algebra of data science is important, and that proper data science is about more than just knowing the math.

Regularization Exercise

Exercise 12-4.

Here you will explore the effects of regularization on the reduced-rank design matrix that you created in the previous exercise. Start by implementing $\left(\mathbf{X}^T\mathbf{X} + \gamma \parallel \mathbf{X} \parallel_F^2 \mathbf{I}\right)^{-1}$, using $\gamma = 0$ and $\gamma = .01$. Print out the size and rank of the two matrices. Here are my results (it's interesting to see that the rank-3 design matrix is so numerically unstable that its "inverse" is actually rank-2):

```
inv(X'X + 0.0*I) size: (4, 4)
inv(X'X + 0.0*I) rank: 2

inv(X'X + 0.01*I) size: (4, 4)
inv(X'X + 0.01*I) rank: 4
```

Now for the experiment. The goal here is to explore the effects of regularization on the fit of the model to the data. Write code that will compute the fit to the data as R^2 using least squares with regularization on the design matrices with and without multicollinearity. Put that code into a for loop that implements a range of γ values between 0 and .2. Then show the results in a figure like Figure 12-10.

By the way, it is trivial that the model fit decreases with increasing regularization for full-rank design matrices—indeed, the purpose of regularization is to make the model less sensitive to the data. The important question is whether regularization improves the fit to a test dataset or validation fold that was excluded when fitting the model. If the regularization is beneficial, you would expect the generalizability of the regularized model to increase up to some γ and then decrease again. That's a level of detail you'd learn about in a dedicated statistics or machine learning book, although you will get to code cross-validation in Chapter 15.

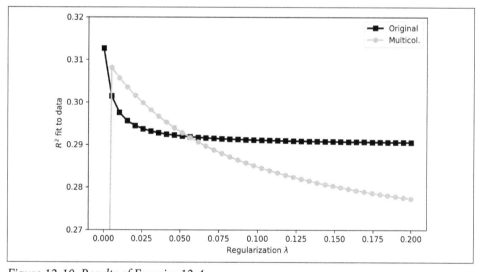

Figure 12-10. Results of Exercise 12-4

Polynomial Regression Exercise

Exercise 12-5.

The purpose of this exercise is to fit the polynomial regression using a range of orders, from zero to nine. In a for loop, recompute the regression and the predicted data values. Show the results like in Figure 12-11.

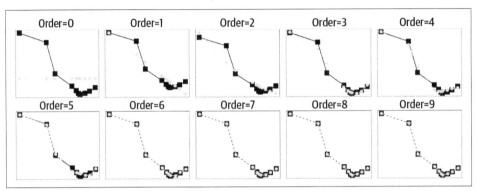

Figure 12-11. Results of Exercise 12-5

This exercise highlights the problems with underfitting and overfitting. The model with too few parameters does a poor job at predicting the data. On the other hand, the model with many parameters fits the data *too well* and risks being overly sensitive to noise and failing to generalize to new data. Strategies for finding the balance between under- and overfitting include cross-validation and Bayes information criterion; these are topics that you would learn about in a machine learning or statistics book.

Grid Search Exercises

Exercise 12-6.

Your goal here is simple: reproduce Figure 12-8 following the instructions presented in the text around that figure. Print out the regression coefficients to compare. For example, the following results were obtained using the grid resolution parameter set to 50:

```
Analytic result:
    Intercept: 23.13, slope: 3.70

Empirical result:
    Intercept: 22.86, slope: 3.67
```

Once you have working code, try a few different resolution parameters. I made Figure 12-8 using a resolution of 100; you should also try other values, e.g., 20 or

500. Also note the computation time for higher resolution values—and this is only a two-parameter model! An exhaustive high-resolution grid search for a 10-parameter model is extremely computationally intensive.

Exercise 12-7.

You have seen two different methods for evaluating the fit of a model to data: sum of squared errors and R^2. In the previous exercise, you used the sum of squared errors to evaluate the model fit to data; in this exercise, you will determine whether R^2 is equally viable.

The coding part of this exercise is simple: modify the code from the previous exercise to compute R^2 instead of SSE (make sure to modify a copy of the code instead of overwriting the previous exercise).

Now for the challenging part: you will find that R^2 is terrible! It gives a completely wrong answer. Your job is to figure out why that is (the online code solution contains a discussion on this point). Hint: store the predicted data from each parameter pair so you can inspect the predicted values, and then compare against the observed data.

## CHAPTER 13

Eigendecomposition

Eigendecomposition is a pearl of linear algebra. What is a pearl? Let me quote directly from the book *20,000 Leagues Under the Sea*:

> For poets, a pearl is a tear from the sea; for Orientals, it's a drop of solidified dew; for the ladies it's a jewel they can wear on their fingers, necks, and ears that's oblong in shape, glassy in luster, and formed from mother-of-pearl; for chemists, it's a mixture of calcium phosphate and calcium carbonate with a little gelatin protein; and finally, for naturalists, it's a simple festering secretion from the organ that produces mother-of-pearl in certain bivalves.
>
> —Jules Verne

The point is that the same object can be seen in different ways depending on its use. So it is with eigendecomposition: eigendecomposition has a geometric interpretation (axes of rotational invariance), a statistical interpetation (directions of maximal covariance), a dynamical-systems interpretation (stable system states), a graph-theoretic interpretation (the impact of a node on its network), a financial-market interpretation (identifying stocks that covary), and many more.

Eigendecomposition (and the SVD, which, as you'll learn in the next chapter, is closely related to eigendecomposition) is among the most important contributions of linear algebra to data science. The purpose of this chapter is to provide you an intuitive understanding of eigenvalues and eigenvectors—the results of eigendecomposition of a matrix. Along the way, you'll learn about diagonalization and more special properties of symmetric matrices. After extending eigendecomposition to the SVD in Chapter 14, you'll see a few applications of eigendecomposition in Chapter 15.

Interpretations of Eigenvalues and Eigenvectors

There are several ways of interpreting eigenvalues/vectors that I will describe in the next sections. Of course, the math is the same regardless, but having mutliple perspectives can facilitate intuition, which in turn will help you understand how and why eigendecomposition is important in data science.

Geometry

I've actually already introduced you to the geometric concept of eigenvectors in Chapter 5. In Figure 5-5, we discovered that there was a special combination of a matrix and a vector such that the matrix *stretched*—but did not *rotate*—that vector. That vector is an eigenvector of the matrix, and the amount of stretching is the eigenvalue.

Figure 13-1 shows vectors before and after multiplication by a 2×2 matrix. The two vectors in the left plot ($\mathbf{v}_1$ and $\mathbf{v}_2$) are eigenvectors whereas the two vectors in the right plot are not. The eigenvectors point in the same direction before and after postmultiplying the matrix. The eigenvalues encode the amount of stretching; try to guess the eigenvalues based on visual inspection of the plot. The answers are in the footnote.[1]

That's the geometric picture: an eigenvector means that matrix-vector multiplication acts like scalar-vector multiplication. Let's see if we can write that down in an equation (we can, and it's printed in Equation 13-1).

Equation 13-1. Eigenvalue equation

$$\mathbf{A}\mathbf{v} = \lambda\mathbf{v}$$

---

1 Approximately –.6 and 1.6.

Be careful with interpreting that equation: it does not say that the matrix *equals* the scalar; it says that the *effect* of the matrix on the vector is the same as the *effect* of the scalar on that same vector.

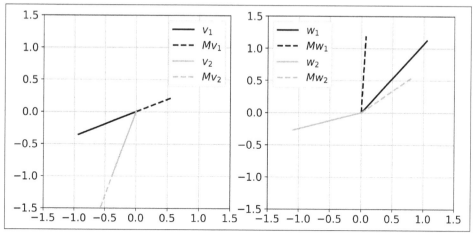

Figure 13-1. Geometry of eigenvectors

This is called the *eigenvalue equation*, and it's another key formula in linear algebra that is worth memorizing. You'll see it throughout this chapter, you'll see a slight variation of it in the following chapter, and you'll see it many times when learning about multivariate statistics, signal processing, optimization, graph theory, and a myriad of other applications where patterns are identified across multiple simultaneously recorded features.

Statistics (Principal Components Analysis)

One of the reasons why people apply statistics is to identify and quantify relationships between variables. For example, the rise of global temperatures correlates with the decline in the number of pirates,[2] but how strong is that relationship? Of course, when you have only two variables, a simple correlation (like what you learned in Chapter 4) is sufficient. But in a multivariate dataset that includes dozens or hundreds of variables, bivariate correlations cannot reveal global patterns.

Let's make this more concrete with an example. Cryptocurrencies are digital stores of value that are encoded in a blockchain, which is a system for keeping track of transactions. You've probably heard of Bitcoin and Ethereum; there are tens of thousands of other cryptocoins that have various purposes. We can ask whether the entirety of the cryptospace operates as a single system (meaning that the value of all coins go

---

2 "Open Letter to Kansas School Board," Church of the Flying Spaghetti Monster, *spaghettimonster.org/about/open-letter*.

up and down together), or whether there are independent subcategories within that space (meaning that some coins or groups of coins change in value independently of the value of other coins).

We can test this hypothesis by performing a principal components analysis on a dataset that contains the prices of various cryptocoins over time. If the entire cryptomarket operates as a single entity, then the *scree plot* (a graph of the eigenvalues of the dataset's covariance matrix) would reveal that one component accounts for most of the variance of the system, and all other components account for very little variance (graph A in Figure 13-2). In contrast, if the cryptomarket had, say, three major subcategories with independent price movements, then we would expect to see three large eigenvalues (graph B in Figure 13-2).

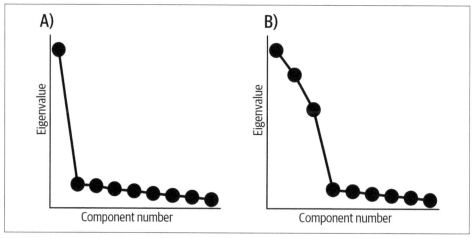

Figure 13-2. Simulated scree plots of multivariate datasets (data is simulated to illustrate outcome possibilities)

Noise Reduction

Most datasets contain noise. *Noise* refers to variance in a dataset that is either unexplained (e.g., random variation) or unwanted (e.g., electrical line noise artifacts in radio signals). There are many ways to attenuate or eliminate noise, and the optimal noise-reduction strategy depends on the nature and origin of the noise and on the characteristics of the signal.

One method of reducing random noise is to identify the eigenvalues and eigenvectors of a system, and "project out" directions in the data space associated with small eigenvalues. The assumption is that random noise makes a relatively small contribution to the total variance. "Projecting out" a data dimension means to reconstruct a dataset after setting some eigenvalues that are below some threshold to zero.

You'll see an example of using eigendecomposition to reduce noise in Chapter 15.

Dimension Reduction (Data Compression)

Information communications technologies like phones, internet, and TV create and transmit a huge amount of data, such as pictures and videos. Transmitting data can be time-consuming and expensive, and it is beneficial to *compress* the data before transmitting it. Compression means to reduce the size of the data (in terms of bytes) while having minimal impact on the quality of the data. For example, a TIFF format image file might be 10 MB, while the JPG converted version might be .1 MB while still retaining reasonably good quality.

One way to dimension-reduce a dataset is to take its eigendecomposition, drop the eigenvalues and eigenvectors associated with small directions in the data space, and then transmit only the relatively larger eigenvector/value pairs. It is actually more common to use the SVD for data compression (and you'll see an example in Chapter 15), although the principle is the same.

Modern data compression algorithms are actually faster and more efficient than the method previously described, but the idea is the same: decompose the dataset into a set of basis vectors that capture the most important features of the data, and then reconstruct a high-quality version of the original data.

Finding Eigenvalues

To eigendecompose a square matrix, you first find the eigenvalues, and then use each eigenvalue to find its corresponding eigenvector. The eigenvalues are like keys that you insert into the matrix to unlock the mystical eigenvector.

Finding the eigenvalues of a matrix is super easy in Python:

```
matrix = np.array([
          [1,2],
          [3,4]
          ])

# get the eigenvalues
evals = np.linalg.eig(matrix)[0]
```

The two eigenvalues (rounded to the nearest hundredth) are −0.37 and 5.37.

But the important question isn't *which function returns the eigenvalues*; instead, the important question is *how are the eigenvalues of a matrix identified?*

To find the eigenvalues of a matrix, we start with the eigenvalue equation shown in Equation 13-1 and do some simple arithemetic, as shown in Equation 13-2.

Equation 13-2. Eigenvalue equation, reorganized

$$\mathbf{Av} = \lambda\mathbf{v}$$
$$\mathbf{Av} - \lambda\mathbf{v} = \mathbf{0}$$
$$(\mathbf{A} - \lambda\mathbf{I})\mathbf{v} = \mathbf{0}$$

The first equation is an exact repeat of the eigenvalue equation. In the second equation, we simply subtracted the right-hand side to set the equation to the zeros vector.

The transition from the second to the third equation requires some explanation. The left-hand side of the second equation has two vector terms, both of which involve $\mathbf{v}$. So we factor out the vector. But that leaves us with the subtraction of a matrix and a scalar $(\mathbf{A} - \lambda)$, which is not a defined operation in linear algebra.[3] So instead, we *shift* the matrix by λ. That brings us to the third equation. (Side note: the expression $\lambda\mathbf{I}$ is sometimes called a *scalar matrix*.)

What does that third equation mean? It means that *the eigenvector is in the null space of the matrix shifted by its eigenvalue*.

If it helps you understand the concept of the eigenvector as the null-space vector of the shifted matrix, you can think of adding two additional equations:

$$\widetilde{\mathbf{A}} = \mathbf{A} - \lambda\mathbf{I}$$
$$\widetilde{\mathbf{A}}\mathbf{v} = \mathbf{0}$$

Why is that statement so insightful? Remember that we ignore trivial solutions in linear algebra, so we do not consider $\mathbf{v} = \mathbf{0}$ to be an eigenvector. And that means that the matrix shifted by its eigenvalue is singular, because only singular matrices have a nontrivial null space.

And what else do we know about singular matrices? We know that their determinant is zero. Hence:

$$|\mathbf{A} - \lambda\mathbf{I}| = 0$$

---

3 As I wrote in Chapter 5, Python will return a result, but that is the scalar broadcast subtracted, which is not a linear algebra operation.

Believe it or not, that's the key to finding eigenvalues: shift the matrix by the unknown eigenvalue λ, set its determinant to zero, and solve for λ. Let's see how this looks for a 2×2 matrix:

$$\left\| \begin{bmatrix} a & b \\ c & d \end{bmatrix} - \lambda \begin{bmatrix} 1 & 0 \\ 0 & 1 \end{bmatrix} \right\| = 0$$

$$\begin{vmatrix} a - \lambda & b \\ c & d - \lambda \end{vmatrix} = 0$$

$$(a - \lambda)(d - \lambda) - bc = 0$$

$$\lambda^2 - (a + d)\lambda + (ad - bc) = 0$$

You could apply the quadratic formula to solve for the two λ values. But the answer itself isn't important; the important thing is to see the logical progression of mathematical concepts established earlier in this book:

- The matrix-vector multiplication acts like scalar-vector multiplication (the eigenvalue equation).
- We set the eigenvalue equation to the zeros vector, and factor out common terms.
- This reveals that the eigenvector is in the null space of the matrix shifted by the eigenvalue. We do not consider the zeros vector to be an eigenvector, which means the shifted matrix is singular.
- Therefore, we set the determinant of the shifted matrix to zero and solve for the unknown eigenvalue.

The determinant of an eigenvalue-shifted matrix set to zero is called the *characteristic polynomial* of the matrix.

Notice that in the previous example, we started with a 2×2 matrix and got a λ^2 term, which means this is a second-order polynomial equation. You might remember from your high-school algebra class that an *n*th order polynomial has *n* solutions, some of which might be complex-valued (this is called the fundamental theorem of algebra). So, there will be two values of λ that can satisfy the equation.

The matching 2s are no coincidence: the characteristic polynomial of an $M \times M$ matrix will have a λ^M term. That is the reason why an $M \times M$ matrix will have M eigenvalues.

Tedious Practice Problems

At this point in a traditional linear algebra textbook, you would be tasked with finding the eigenvalues of dozens of 2×2 and 3×3 matrices by hand. I have mixed feelings about these kinds of exercises: on the one hand, solving problems by hand really helps internalize the mechanics of finding eigenvalues; but on the other hand, I want to focus this book on concepts, code, and applications, without getting bogged down by tedious arithmetic. If you feel inspired to solve eigenvalue problems by hand, then go for it! You can find myriad such problems in traditional textbooks or online. But I took the bold (and perhaps controversial) decision to avoid hand-solved problems in this book, and instead to have exercises that focus on coding and comprehension.

Finding Eigenvectors

As with eigenvalues, finding eigenvectors is super-duper easy in Python:

```
evals,evecs = np.linalg.eig(matrix)
print(evals), print(evecs)

  [-0.37228132  5.37228132]

  [[-0.82456484 -0.41597356]
   [ 0.56576746 -0.90937671]]
```

The eigenvectors are in the columns of the matrix evecs and are in the same order as the eigenvalues (that is, the eigenvector in the first column of matrix evecs is paired with the first eigenvalue in vector evals). I like to use the variable names evals and evecs, because they are short and meaningful. You might also see people use variable names L and V or D and V. The L is for Λ (the capital of λ) and the V is for **V**, the matrix in which each column i is eigenvector $\mathbf{v}_i$. The D is for *diagonal*, because eigenvalues are often stored in a diagonal matrix, for reasons I will explain later in this chapter.

Eigenvectors in the Columns, Not the Rows!

The most important thing to keep in mind about eigenvectors when coding is that they are stored in the *columns of the matrix*, not the rows! Such dimensional-indexing errors are easy to make with square matrices (because you might not get Python errors), but accidentally using the rows instead of the columns of the eigenvectors matrix can have disastrous consequences in applications. When in doubt, remember the discussion from Chapter 2 that common convention in linear algebra is to assume that vectors are in column orientation.

OK, but again, the previous code shows how to get a NumPy function to return the eigenvectors of a matrix. You could have learned that from the `np.linalg.eig` docstring. The important question is *Where do eigenvectors come from, and how do we find them?*

Actually, I've already written how to find eigenvectors: find the vector $\mathbf{v}$ that is in the null space of the matrix shifted by λ. In other words:

$$\mathbf{v}_i \in N(\mathbf{A} - \lambda_i\mathbf{I})$$

Let's see a numerical example. Follwing is a matrix and its eigenvalues:

$$\begin{bmatrix} 1 & 2 \\ 2 & 1 \end{bmatrix} \implies \lambda_1 = 3,\ \lambda_2 = -1$$

Let's focus on the first eigenvalue. To reveal its eigenvector, we shift the matrix by 3 and find a vector in its null space:

$$\begin{bmatrix} 1-3 & 2 \\ 2 & 1-3 \end{bmatrix} = \begin{bmatrix} -2 & 2 \\ 2 & -2 \end{bmatrix} \implies \begin{bmatrix} -2 & 2 \\ 2 & -2 \end{bmatrix}\begin{bmatrix} 1 \\ 1 \end{bmatrix} = \begin{bmatrix} 0 \\ 0 \end{bmatrix}$$

This means that [1 1] is an eigenvector of the matrix associated with an eigenvalue of 3.

I found that null space vector just by looking at the matrix. How are the null space vectors (that is, the eigenvectors of the matrix) actually identified in practice?

Null space vectors can be found by using Gauss-Jordan to solve a system of equations, where the coefficients matrix is the λ-shifted matrix and the constants vector is the zeros vector. That's a good way to conceptualize the solution. In implementation, more stable numerical methods are applied for finding eigenvalues and eigenvectors, including QR decomposition and a procedure called the power method.

Sign and Scale Indeterminacy of Eigenvectors

Let me return to the numerical example in the previous section. I wrote that [1 1] was an eigenvector of the matrix because that vector is a basis for the null space of the matrix shifted by its eigenvalue of 3.

Look back at the shifted matrix and ask yourself, is [1 1] the only possible basis vector for the null space? Not even close! You could also use [4 4] or [−5.4 −5.4] or...I think you see where this is going: *any* scaled version of vector [1 1] is a basis for that null space. In other words, if $\mathbf{v}$ is an eigenvector of a matrix, then so is $\alpha\mathbf{v}$ for any real-valued α except zero.

Indeed, eigenvectors are important because of their *direction*, not because of their *magnitude*.

The infinity of possible null space basis vectors leads to two questions:

- *Is there one "best" basis vector?* There isn't a "best" basis vector per se, but it is convenient to have eigenvectors that are unit normalized (a Euclidean norm of 1). This is particularly useful for symmetric matrices for reasons that will be explained later in this chapter.[4]

- *What is the "correct" sign of an eigenvector?* There is none. In fact, you can get different eigenvector signs from the same matrix when using different versions of NumPy—as well as different software such as MATLAB, Julia, or Mathematica. Eigenvector sign indeterminacy is just a feature of life in our universe. In applications such as PCA, there are principled ways for assigning a sign, but that's just common convention to facilitate interpretation.

Diagonalizing a Square Matrix

The eigenvalue equation that you are now familiar with lists one eigenvalue and one eigenvector. This means that an $M \times M$ matrix has M eigenvalue equations:

$$\mathbf{A}\mathbf{v}_1 = \lambda_1 \mathbf{v}_1$$
$$\vdots$$
$$\mathbf{A}\mathbf{v}_M = \lambda_M \mathbf{v}_M$$

There's nothing really wrong with that series of equations, but it is kind of ugly, and ugliness violates one of the principles of linear algebra: make equations compact and elegant. Therefore, we transform this series of equations into one matrix equation.

The key insight for writing out the matrix eigenvalue equation is that each column of the eigenvectors matrix is scaled by exactly one eigenvalue. We can implement this through postmultiplication by a diagonal matrix (as you learned in Chapter 6).

So instead of storing the eigenvalues in a vector, we store the eigenvalues in the diagonal of a matrix. The following equation shows the form of diagonalization for a 3×3 matrix (using @ in place of numerical values in the matrix). In the eigenvectors matrix, the first subscript number corresponds to the eigenvector, and the second subscript number corresponds to the eigenvector element. For example, v_{12} is the second element of the first eigenvector:

---

4 To quell suspense: it makes the eigenvectors matrix an orthogonal matrix.

$$
\begin{bmatrix} @ & @ & @ \\ @ & @ & @ \\ @ & @ & @ \end{bmatrix}
\begin{bmatrix} v_{11} & v_{21} & v_{31} \\ v_{12} & v_{22} & v_{32} \\ v_{13} & v_{23} & v_{33} \end{bmatrix}
=
\begin{bmatrix} v_{11} & v_{21} & v_{31} \\ v_{12} & v_{22} & v_{32} \\ v_{13} & v_{23} & v_{33} \end{bmatrix}
\begin{bmatrix} \lambda_1 & 0 & 0 \\ 0 & \lambda_2 & 0 \\ 0 & 0 & \lambda_3 \end{bmatrix}
$$

$$
=
\begin{bmatrix} \lambda_1 v_{11} & \lambda_2 v_{21} & \lambda_3 v_{31} \\ \lambda_1 v_{12} & \lambda_2 v_{22} & \lambda_3 v_{32} \\ \lambda_1 v_{13} & \lambda_2 v_{23} & \lambda_3 v_{33} \end{bmatrix}
$$

Please take a moment to confirm that each eigenvalue scales all elements of its corresponding eigenvector, and not any other eigenvectors.

More generally, the matrix eigenvalue equation—a.k.a. the diagonalization of a square matrix—is:

$$\mathbf{AV} = \mathbf{V\Lambda}$$

NumPy's `eig` function returns eigenvectors in a matrix and eigenvalues in a vector. This means that diagonalizing a matrix in NumPy requires a bit of extra code:

```
evals,evecs = np.linalg.eig(matrix)
D = np.diag(evals)
```

By the way, it's often interesting and insightful in mathematics to rearrange equations by solving for different variables. Consider the following list of equivalent declarations:

$$\mathbf{AV} = \mathbf{V\Lambda}$$
$$\mathbf{A} = \mathbf{V\Lambda V}^{-1}$$
$$\mathbf{\Lambda} = \mathbf{V}^{-1}\mathbf{AV}$$

The second equation shows that matrix $\mathbf{A}$ becomes diagonal inside the space of $\mathbf{V}$ (that is, $\mathbf{V}$ moves us into the "diagonal space," and then $\mathbf{V}^{-1}$ gets us back out of the diagonal space). This can be interpreted in the context of basis vectors: the matrix $\mathbf{A}$ is dense in the standard basis, but then we apply a set of transformations ($\mathbf{V}$) to rotate the matrix into a new set of basis vectors (the eigenvectors) in which the information is sparse and represented by a diagonal matrix. (At the end of the equation, we need to get back into the standard basis space, hence the $\mathbf{V}^{-1}$.)

The Special Awesomeness of Symmetric Matrices

You already know from earlier chapters that symmetric matrices have special properties that make them great to work with. Now you are ready to learn two more special properties that relate to eigendecomposition.

Orthogonal Eigenvectors

Symmetric matrices have orthogonal eigenvectors. That means that all eigenvectors of a symmetric matrix are pair-wise orthogonal. Let me start with an example, then I'll discuss the implications of eigenvector orthogonality, and finally I'll show the proof:

```
# just some random symmetric matrix
A = np.random.randint(-3,4,(3,3))
A = A.T@A

# its eigendecomposition
L,V = np.linalg.eig(A)

# all pairwise dot products
print( np.dot(V[:,0],V[:,1]) )
print( np.dot(V[:,0],V[:,2]) )
print( np.dot(V[:,1],V[:,2]) )
```

The three dot products are all zero (within computer rounding errors on the order of 10^{-16}. (Notice that I've created symmetric matrices as a random matrix times its transpose.)

The orthogonal eigenvector property means that the dot product between any pair of eigenvectors is zero, while the dot product of an eigenvector with itself is nonzero (because we do not consider the zeros vector to be an eigenvector). This can be written as $\mathbf{V}^\mathsf{T}\mathbf{V} = \mathbf{D}$, where $\mathbf{D}$ is a diagonal matrix with the diagonals containing the norms of the eigenvectors.

But we can do even better than just a diagonal matrix: recall that eigenvectors are important because of their *direction*, not because of their *magnitude*. So an eigenvector can have any magnitude we want (except, obviously, for a magnitude of zero).

Let's scale all eigenvectors so they have unit length. Question for you: if all eigenvectors are orthogonal and have unit length, what happens when we multiply the eigenvectors matrix by its transpose?

Of course you know the answer:

$$\mathbf{V}^\mathsf{T}\mathbf{V} = \mathbf{I}$$

In other words, the eigenvectors matrix of a symmetric matrix is an orthogonal matrix! This has multiple implications for data science, including that the eigenvectors are super easy to invert (because you simply transpose them). There are other implications of orthogonal eigenvectors for applications such as principal components analysis, which I will discuss in Chapter 15.

I wrote in Chapter 1 that there are relatively few proofs in this book. But orthogonal eigenvectors of symmetric matrices is such an important concept that you really need to see this claim proven.

The goal of this proof is to show that the dot product between any pair of eigenvectors is zero. We start from two assumptions: (1) matrix $\mathbf{A}$ is symmetric, and (2) λ_1 and λ_2 are distinct eigenvalues of $\mathbf{A}$ (*distinct* meaning they cannot equal each other), with $\mathbf{v}_1$ and $\mathbf{v}_2$ as their corresponding eigenvectors. Try to follow each equality step from left to right of Equation 13-3.

Equation 13-3. Proof of eigenvector orthogonality for symmetric matrices

$$\lambda_1 \mathbf{v}_1^T \mathbf{v}_2 = (\mathbf{A}\mathbf{v}_1)^T \mathbf{v}_2 = \mathbf{v}_1^T \mathbf{A}^T \mathbf{v}_2 = \mathbf{v}_1^T \lambda_2 \mathbf{v}_2 = \lambda_2 \mathbf{v}_1^T \mathbf{v}_2$$

The terms in the middle are just transformations; pay attention to the first and last terms. They are rewritten in Equation 13-4, and then subtracted to set to zero.

Equation 13-4. Continuing the eigenvector orthogonality proof...

$$\lambda_1 \mathbf{v}_1^T \mathbf{v}_2 = \lambda_2 \mathbf{v}_1^T \mathbf{v}_2$$
$$\lambda_1 \mathbf{v}_1^T \mathbf{v}_2 - \lambda_2 \mathbf{v}_1^T \mathbf{v}_2 = 0$$

Both terms contain the dot product $\mathbf{v}_1^T \mathbf{v}_2$, which can be factored out. This brings us to the final part of the proof, which is shown in Equation 13-5.

Equation 13-5. Eigenvector orthogonality proof, part 3

$$(\lambda_1 - \lambda_2)\mathbf{v}_1^T \mathbf{v}_2 = 0$$

This final equation says that two quantities multiply to produce 0, which means that one or both of those quantities must be zero. $(\lambda_1 - \lambda_2)$ cannot equal zero because we began from the assumption that they are distinct. Therefore, $\mathbf{v}_1^T \mathbf{v}_2$ must equal zero, which means that the two eigenvectors are orthogonal. Go back through the equations to convince yourself that this proof fails for nonsymmetric matrices, when $\mathbf{A}^T \neq \mathbf{A}$. Thus, the eigenvectors of a nonsymmetric matrix are not constrained to be orthogonal (they will be linearly independent for all distinct eigenvalues, but I will omit that discussion and proof).

Real-Valued Eigenvalues

A second special property of symmetric matrices is that they have real-valued eigenvalues (and therefore real-valued eigenvectors).

Let me start by showing that matrices—even with all real-valued entries—can have complex-valued eigenvalues:

```
A = np.array([[-3, -3, 0],
              [ 3, -2, 3],
              [ 0,  1, 2]])

# its eigendecomposition
L,V = np.linalg.eig(A)
L.reshape(-1,1) # print as column vector

>> array([[-2.744739 +2.85172624j],
          [-2.744739 -2.85172624j],
          [ 2.489478 +0.j        ]])
```

(Be careful with interpreting that NumPy array; it is not a 3×2 *matrix*; it is a 3×1 *column vector* that contains complex numbers. Note the j and the absence of a comma between numbers.)

The 3×3 matrix **A** has two complex eigenvalues and one real-valued eigenvalue. The eigenvectors coupled to the complex-valued eigenvalues will themselves be complex-valued. There is nothing special about that particular matrix; I literally generated it from random integers between -3 and $+3$. Interestingly, complex-valued solutions come in conjugate pairs. That means that if there is a $\lambda_j = a + ib$, then there is a $\lambda_k = a - ib$. Their corresponding eigenvectors are also complex conjugate pairs.

I don't want to go into detail about complex-valued solutions, except to show you that complex solutions to eigendecomposition are straightforward.[5]

Symmetric matrices are guaranteed to have real-valued eigenvalues, and therefore also real-valued eigenvectors. Let me start by modifying the previous example to make the matrix symmetric:

```
A = np.array([[-3, -3, 0],
              [-3, -2, 1],
              [ 0,  1, 2]])

# its eigendecomposition
```

---

5 By "straightforward" I mean mathematically expected; interpreting complex solutions in eigendecomposition is far from straightforward.

```
L,V = np.linalg.eig(A)
L.reshape(-1,1) # print as column vector

>> array([[-5.59707146],
          [ 0.22606174],
          [ 2.37100972]])
```

This is just one specific example; maybe we got lucky here? I recommend taking a moment to explore this yourself in the online code; you can create random symmetric matrices (by creaing a random matrix and eigendecomposing $\mathbf{A}^T\mathbf{A}$) of any size to confirm that the eigenvalues are real-valued.

Guaranteed real-valued eigenvalues from symmetric matrices is fortunate because complex numbers are often confusing to work with. Lots of matrices used in data science are symmetric, and so if you see complex eigenvalues in your data science applications, it's possible that there is a problem with the code or with the data.

Leveraging Symmetry

If you know that you are working with a symmetric matrix, you can use np.linalg.eigh instead of np.linalg.eig (or SciPy's eigh instead of eig). The h is for "Hermitian," which is the complex version of a symmetric matrix. eigh can be faster and more numerically stable than eig, but works only on symmetric matrices.

Eigendecomposition of Singular Matrices

I included this section here because I find that students often get the idea that singular matrices cannot be eigendecomposed, or that the eigenvectors of a singular matrix must be unusual somehow.

That idea is completely wrong. Eigendecomposition of singular matrices is perfectly fine. Here is a quick example:

```
# a singular matrix
A = np.array([[1,4,7],
              [2,5,8],
              [3,6,9]])

# its eigendecomposition
L,V = np.linalg.eig(A)
```

The rank, eigenvalues, and eigenvectors of this matrix are printed here:

```
print( f'Rank = {np.linalg.matrix_rank(A)}\n' )
print('Eigenvalues: '), print(L.round(2)), print(' ')
print('Eigenvectors:'), print(V.round(2))

>> Rank = 2
```

```
Eigenvalues:
[16.12 -1.12 -0.  ]

Eigenvectors:
[[-0.46 -0.88  0.41]
 [-0.57 -0.24 -0.82]
 [-0.68  0.4   0.41]]
```

This rank-2 matrix has one zero-valued eigenvalue with a nonzeros eigenvector. You can use the online code to explore the eigendecomposition of other reduced-rank random matrices.

There is one special property of the eigendecomposition of singular matrices, which is that at least one eigenvalue is guaranteed to be zero. That doesn't mean that the number of nonzero eigenvalues equals the rank of the matrix—that's true for singular values (the scalar values from the SVD) but not for eigenvalues. But if the matrix is singular, then at least one eigenvalue equals zero.

The converse is also true: every full-rank matrix has zero zero-valued eigenvalues.

One explanation for why this happens is that a singular matrix already has a non-trivial null space, which means $\lambda = 0$ provides a nontrivial solution to the equation $(\mathbf{A} - \lambda\mathbf{I})\mathbf{v} = \mathbf{0}$. You can see this in the previous example matrix: the eigenvector associated with $\lambda = 0$ is the normalized vector [1 −2 1], which is the linear weighted combination of the columns (or rows) that produces the zeros vector.

The main take-homes of this section are (1) eigendecomposition is valid for reduced-rank matrices, and (2) the presence of at least one zero-valued eigenvalue indicates a reduced-rank matrix.

Quadratic Form, Definiteness, and Eigenvalues

Let's face it: *quadratic form* and *definiteness* are intimidating terms. But don't worry—they are both straightforward concepts that provide a gateway to advanced linear algebra and applications such as principal components analysis and Monte Carlo simulations. And better still: integrating Python code into your learning will give you a huge advantage over learning about these concepts compared to traditional linear algebra textbooks.

The Quadratic Form of a Matrix

Consider the following expression:

$$\mathbf{w}^\mathrm{T}\mathbf{A}\mathbf{w} = \alpha$$

In other words, we pre- and postmultiply a square matrix by the same vector **w** and get a scalar. (Notice that this multiplication is valid only for square matrices.)

This is called the *quadratic form* on matrix **A**.

Which matrix and which vector do we use? The idea of the quadratic form is to use one specific matrix and the set of all possible vectors (of appropriate size). The important question concerns the signs of α for all possible vectors. Let's see an example:

$$[x \ y]\begin{bmatrix} 2 & 4 \\ 0 & 3 \end{bmatrix}\begin{bmatrix} x \\ y \end{bmatrix} = 2x^2 + (0+4)xy + 3y^2$$

For this particular matrix, there is no possible combination of x and y that can give a negative answer, because the squared terms ($2x^2$ and $3y^2$) will always overpower the cross-term ($4xy$) even when x or y is negative. Furthermore, α can be nonpositive only when $x = y = 0$.

That is not a trivial result of the quadratic form. For example, the following matrix can have a positive or negative α depending on the values of x and y:

$$[x \ y]\begin{bmatrix} -9 & 4 \\ 3 & 9 \end{bmatrix}\begin{bmatrix} x \\ y \end{bmatrix} = -9x^2 + (3+4)xy + 9y^2$$

You can confirm that setting $[x \ y]$ to $[-1 \ 1]$ gives a negative quadratic form result, while $[-1 \ -1]$ gives a positive result.

How can you possibly know whether the quadratic form will produce a positive (or negative, or zero-valued) scalar for *all* possible vectors? The key comes from considering that a full-rank eigenvectors matrix spans all of $\mathbb{R}^M$, and therefore that every vector in $\mathbb{R}^M$ can be expressed as some linear weighted combination of the eigenvectors.[6] Then we start from the eigenvalue equation and left-multiply by an eigenvector to return to the quadratic form:

$$\mathbf{Av} = \lambda\mathbf{v}$$
$$\mathbf{v}^\mathrm{T}\mathbf{Av} = \lambda\mathbf{v}^\mathrm{T}\mathbf{v}$$
$$\mathbf{v}^\mathrm{T}\mathbf{Av} = \lambda\|\mathbf{v}\|^2$$

---

6 In the interest of brevity, I am omitting some subtlety here about rare cases where the eigenvectors matrix does not span the entire M-dimensional subspace.

The final equation is key. Note that $\| \mathbf{v}^T\mathbf{v} \|$ is strictly positive (vector magnitudes cannot be negative, and we ignore the zeros vector), which means that the sign of the right-hand side of the equation is determined entirely by the eigenvalue λ.

That equation uses only one eigenvalue and its eigenvector, but we need to know about any possible vector. The insight is to consider that if the equation is valid for each eigenvector-eigenvalue pair, then it is valid for any combination of eigenvector-eigenvalue pairs. For example:

$$\mathbf{v}_1^T\mathbf{A}\mathbf{v}_1 = \lambda_1 \| \mathbf{v}_1 \|^2$$
$$\mathbf{v}_2^T\mathbf{A}\mathbf{v}_2 = \lambda_2 \| \mathbf{v}_2 \|^2$$
$$(\mathbf{v}_1 + \mathbf{v}_2)^T\mathbf{A}(\mathbf{v}_1 + \mathbf{v}_2) = (\lambda_1 + \lambda_2) \| (\mathbf{v}_1 + \mathbf{v}_2) \|^2$$
$$\mathbf{u}^T\mathbf{A}\mathbf{u} = \zeta \| \mathbf{u} \|^2$$

In other words, we can set any vector $\mathbf{u}$ to be some linear combination of eigenvectors, and some scalar ζ to be that same linear combination of eigenvalues. Anyway, it doesn't change the principle that the sign of the right-hand side—and therefore also the sign of the quadratic form—is determined by the sign of the eigenvalues.

Now let's think about these equations under different assumptions about the signs of the λs:

All eigenvalues are positive
    The right-hand side of the equation is always positive, meaning that $\mathbf{v}^T\mathbf{A}\mathbf{v}$ is always positive for any vector $\mathbf{v}$.

Eigenvalues are positive or zero
    $\mathbf{v}^T\mathbf{A}\mathbf{v}$ is nonnegative and will equal zero when $\lambda = 0$ (which happens when the matrix is singular).

Eigenvalues are negative or zero
    The quadratic form result will be zero or negative.

Eigenvalues are negative
    The quadratic form result will be negative for all vectors.

Definiteness

Definiteness is a characteristic of a square matrix and is defined by the signs of the eigenvalues of the matrix, which is the same thing as the signs of the quadratic form results. Definiteness has implications for the invertibility of a matrix as well as advanced data analysis methods such as generalized eigendecomposition (used in multivariate linear classifiers and signal processing).

There are five categories of definiteness, as shown in Table 13-1; the + and − signs indicate the signs of the eigenvalues.

Table 13-1. Definiteness categories

Category	Quadratic form	Eigenvalues	Invertible
Positive definite	Positive	+	Yes
Positive semidefinite	Nonnegative	+ and 0	No
Indefinite	Positive and negative	+ and −	Depends
Negative semidefinite	Nonpositive	− and 0	No
Negative definite	Negative	−	Yes

"Depends" in the table means that the matrix can be invertible or singular depending on the numbers in the matrix, not on the definiteness category.

A^TA Is Positive (Semi)definite

Any matrix that can be expressed as the product of a matrix and its transpose (that is, $S = A^TA$) is guaranteed to be positive definite or positive semidefinite. The combination of these two categories is often written as "positive (semi)definite."

All data covariance matrices are positive (semi)definite, because they are defined as the data matrix times its transpose. This means that all covariance matrices have nonnegative eigenvalues. The eigenvalues will be all positive when the data matrix is full-rank (full column-rank if the data is stored as observations by features), and there will be at least one zero-valued eigenvalue if the data matrix is reduced-rank.

The proof that S is positive (semi)definite comes from writing out its quadratic form and applying some algebraic manipulations. (Observe that the transition from the first to the second equation simply involves moving parentheses around; such "proof by parentheses" is common in linear algebra.)

$$\begin{aligned}
\mathbf{w}^T\mathbf{S}\mathbf{w} &= \mathbf{w}^T\left(\mathbf{A}^T\mathbf{A}\right)\mathbf{w} \\
&= \left(\mathbf{w}^T\mathbf{A}^T\right)(\mathbf{A}\mathbf{w}) \\
&= (\mathbf{A}\mathbf{w})^T(\mathbf{A}\mathbf{w}) \\
&= \|\mathbf{A}\mathbf{w}\|^2
\end{aligned}$$

The point is that the quadratic form of A^TA equals the squared magnitude of a matrix times a vector. Magnitudes cannot be negative, and can be zero only when the vector is zero. And if $Aw = 0$ for a nontrival w, then A is singular.

Keep in mind that although all $\mathbf{A}^T\mathbf{A}$ matrices are symmetric, not all symmetric matrices can be expressed as $\mathbf{A}^T\mathbf{A}$. In other words, matrix symmetry on its own does not guarantee positive (semi)definiteness, because not all symmetric matrices can be expressed as the product of a matrix and its transpose.

Who Cares About the Quadratic Form and Definiteness?

Positive definiteness is relevant for data science because there are some linear algebra operations that apply only to such well-endowed matrices, including Cholesky decomposition, used to create correlated datasets in Monte Carlo simulations. Positive definite matrices are also important for optimization problems (e.g., gradient descent), because there is a guaranteed minimum to find. In your never-ending quest to improve your data science prowess, you might encounter technical papers that use the abbreviation SPD: symmetric positive definite.

Generalized Eigendecomposition

Consider that the following equation is the same as the fundamental eigenvalue equation:

$$\mathbf{Av} = \lambda\mathbf{Iv}$$

This is obvious because $\mathbf{Iv} = \mathbf{v}$. Generalized eigendecomposition involves replacing the identity matrix with another matrix (not the identity or zeros matrix):

$$\mathbf{Av} = \lambda\mathbf{Bv}$$

Generalized eigendecomposition is also called *simultaneous diagonalization of two matrices*. The resulting $(\lambda, \mathbf{v})$ pair is not an eigenvalue/vector of $\mathbf{A}$ alone nor of $\mathbf{B}$ alone. Instead, the two matrices share eigenvalue/vector pairs.

Conceptually, you can think of generalized eigendecomposition as the "regular" eigendecomposition of a product matrix:

$$\mathbf{C} = \mathbf{AB}^{-1}$$
$$\mathbf{Cv} = \lambda\mathbf{v}$$

This is just conceptual; in practice, generalized eigendecomposition does not require $\mathbf{B}$ to be invertible.

It is not the case that any two matrices can be simultaneously diagonalized. But this diagonalization is possible if **B** is positive (semi)definite.

NumPy does not compute generalized eigendecomposition, but SciPy does. If you know that the two matrices are symmetric, you can use the function `eigh`, which is more numerically stable:

```
# create correlated matrices
A = np.random.randn(4,4)
A = A@A.T
B = np.random.randn(4,4)
B = B@B.T + A/10

# GED
from scipy.linalg import eigh
evals,evecs = eigh(A,B)
```

Be mindful of the order of inputs: the second input is the one that is conceptually inverted.

In data science, generalized eigendecomposition is used in classification analysis. In particular, Fisher's linear discriminant analysis is based on the generalized eigendecomposition of two data covariance matrices. You'll see an example in Chapter 15.

The Myriad Subtleties of Eigendecomposition

There is a lot more that could be said about the properties of eigendecomposition. A few examples: the sum of the eigenvalues equals the trace of the matrix, while the product of the eigenvalues equals the determinant; not all square matrices can be diagonalized; some matrices have repeated eigenvalues, which has implications for their eigenvectors; complex eigenvalues of real-valued matrices are inside a circle in the complex plane. The mathematical knowledge of eigenvalues runs deep, but this chapter provides the essential foundational knowledge for working with eigendecomposition in applications.

Summary

That was quite a chapter! Here is a reminder of the key points:

- Eigendecomposition identifies M scalar/vector pairs of an $M \times M$ matrix. Those pairs of eigenvalue/eigenvector reflect special directions in the matrix and have myriad applications in data science (principal components analysis being a common one), as well as in geometry, physics, computational biology, and myriad other technical disciplines.

- Eigenvalues are found by assuming that the matrix shifted by an unknown scalar λ is singular, setting its determinant to zero (called the *characteristic polynomial*), and solving for the λs.

- Eigenvectors are found by finding the basis vector for the null space of the λ-shifted matrix.

- *Diagonalizing a matrix* means to represent the matrix as $\mathbf{V}^{-1}\boldsymbol{\Lambda}\mathbf{V}$, where $\mathbf{V}$ is a matrix with eigenvectors in the columns and $\boldsymbol{\Lambda}$ is a diagonal matrix with eigenvalues in the diagonal elements.

- Symmetric matrices have several special properties in eigendecomposition; the most relevant for data science is that all eigenvectors are pair-wise orthogonal. This means that the matrix of eigenvectors is an orthogonal matrix (when the eigenvectors are unit normalized), which in turn means that the inverse of the eigenvectors matrix is its transpose.

- The *definiteness* of a matrix refers to the signs of its eigenvalues. In data science, the most relevant categories are positive (semi)definite, which means that all eigenvalues are either nonnegative or positive.

- A matrix times its transpose is always positive (semi)definite, which means all covariance matrices have nonnegative eigenvalues.

- The study of eigendecomposition is rich and detailed, and many fascinating subtleties, special cases, and applications have been discovered. I hope that the overview in this chapter provides a solid grounding for your needs as a data scientist, and may have inspired you to learn more about the fantastic beauty of eigendecomposition.

Code Exercises

Exercise 13-1.

Interestingly, the eigenvectors of $\mathbf{A}^{-1}$ are the same as the eigenvectors of $\mathbf{A}$ while the eigenvalues are λ^{-1}. Prove that this is the case by writing out the eigendecomposition of $\mathbf{A}$ and $\mathbf{A}^{-1}$. Then illustrate it using a random full-rank 5×5 symmetric matrix.

Exercise 13-2.

Re-create the left-side panel of Figure 13-1, but using the *rows* of $\mathbf{V}$ instead of *columns*. Of course you know that this is a coding error, but the results are insightful: it fails the geometry test that the matrix times its eigenvector only stretches.

Exercise 13-3.

The goal of this exercise is to demonstrate that eigenvalues are inextricably coupled to their eigenvectors. Diagonalize a symmetric random-integers matrix[7] created using the additive method (see Exercise 5-9), but randomly reorder the eigenvalues (let's call this matrix $\widetilde{\Lambda}$) without reordering the eigenvectors.

First, demonstrate that you can reconstruct the original matrix as $\mathbf{V\Lambda V}^{-1}$. You can compute reconstruction accuracy as the Frobenius distance between the original and reconstructed matrix. Next, attempt to reconstruct the matrix using $\widetilde{\Lambda}$. How close is the reconstructed matrix to the original? What happens if you only swap the two largest eigenvalues instead of randomly reordering them? How about the two smallest eigenvalues?

Finally, create a bar plot showing the Frobenius distances to the original matrix for the different swapping options (Figure 13-3). (Of course, because of the random matrices—and thus, random eigenvalues—your plot won't look exactly like mine.)

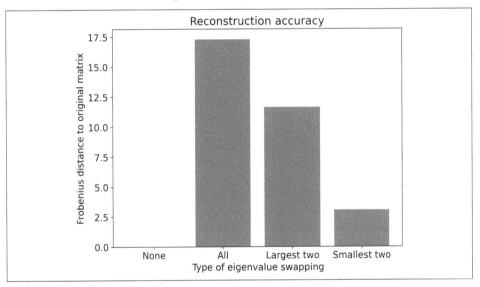

Figure 13-3. Results of Exercise 13-3

Exercise 13-4.

One interesting property of random matrices is that their complex-valued eigenvalues are distributed in a circle with a radius proportional to the size of the matrix. To

---

7 I often use symmetric matrices in exercises because they have real-valued eigenvalues, but that doesn't change the principle or the math; it merely facilitates visual inspection of the solutions.

demonstrate this, compute 123 random 42×42 matrices, extract their eigenvalues, divide by the square root of the matrix size (42), and plot the eigenvalues on the complex plane, as in Figure 13-4.

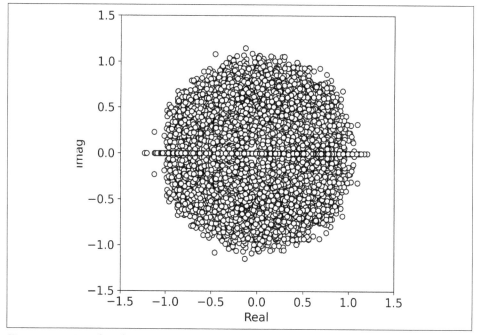

Figure 13-4. Results of Exercise 13-4

Exercise 13-5.

This exercise will help you better understand that an eigenvector is the basis for the null space of the eigenvalue-shifted matrix—and it will also reveal the risks of numerical precision errors. Eigendecompose a random 3×3 symmetric matrix. Then for each eigenvalue, use `scipy.linalg.null_space()` to find a basis vector for the null space of each shifted matrix. Are those vectors the same as the eigenvectors? Note that you might need to take into consideration the norms and the sign indeterminacies of eigenvectors.

When you run the code multiple times for different random matrices, you are likely to get Python errors. The error comes from an empty null space for the λ-shifted matrix, which, upon investigation, comes from the shifted matrix being full rank. (Don't take my word for it; confirm this yourself!) That is not supposed to happen, which highlights—yet again—that (1) finite-precision math on computers does not always conform to chalkboard math and (2) you should use the targeted and more numerically stable functions instead of trying to make direct translations of formulas into code.

Exercise 13-6.

I'm going to teach you a third method to create random symmetric matrices.[8] Start by creating a 4×4 diagonal matrix with positive numbers on the diagonals (they can be, for example, the numbers 1, 2, 3, 4). Then create a 4×4 **Q** matrix from the QR decomposition of a random-numbers matrix. Use these matrices as the eigenvalues and eigenvectors, and multiply them appropriately to assemble a matrix. Confirm that the assembled matrix is symmetric, and that its eigenvalues equal the eigenvalues you specified.

Exercise 13-7.

Let's revisit Exercise 12-4. Redo that exercise but use the average of the eigenvalues instead of the squared Frobenius norm of the design matrix (this is known as *shrinkage regularization*). How does the resulting figure compare with that from Chapter 12?

Exercise 13-8.

This and the following exercise are closely linked. We will create surrogate data with a specified correlation matrix (this exercise), and then remove the correlation (next exercise). The formula to create data with a specified correlation structure is:

$$Y = V\Lambda^{1/2}X$$

where **V** and **Λ** are the eigenvectors and eigenvalues of a correlation matrix, and **X** is an $N \times T$ matrix of uncorrelated random numbers (N features and T observations).

Apply that formula to create a $3 \times 10{,}000$ data matrix **Y** with the following correlation structure:

$$R = \begin{bmatrix} 1 & .2 & .9 \\ .2 & 1 & .3 \\ .9 & .3 & 1 \end{bmatrix}$$

Then compute the empirical correlation matrix of the data matrix **Y**. It won't exactly equal **R** because we are randomly sampling a finite dataset. But it should be fairly close (e.g., within .01).

---

8 The first two were the multiplicative and additive methods.

Exercise 13-9.

Now let's remove those imposed correlations by *whitening*. Whitening is a term in signal and image processing to remove correlations. A multivariate time series can be whitened by implementing the following formula:

$$\widetilde{\mathbf{Y}} = \mathbf{Y}^T\mathbf{V}\mathbf{\Lambda}^{-1/2}$$

Apply that formula to the data matrix from the previous exercise, and confirm that the correlation matrix is the identity matrix (again, within some tolerance for random sampling).

Exercise 13-10.

In generalized eigendecomposition, the eigenvectors are not orthogonal, even when both matrices are symmetric. Confirm in Python that $\mathbf{V}^{-1} \neq \mathbf{V}^T$. This happens because although both $\mathbf{A}$ and $\mathbf{B}$ are symmetric, $\mathbf{C} = \mathbf{A}\mathbf{B}$ is not symmetric.[9]

However, the eigenvectors are orthogonal with respect to $\mathbf{B}$, which means that $\mathbf{V}^T\mathbf{B}\mathbf{V} = \mathbf{I}$. Confirm these properties by performing a generalized eigendecomposition on two symmetric matrices, and producing Figure 13-5.

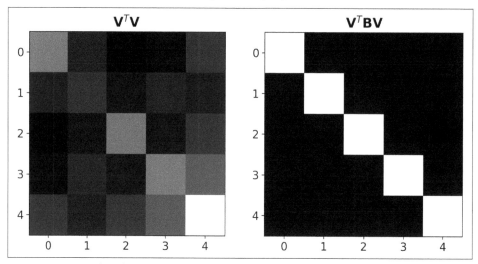

Figure 13-5. Results of Exercise 13-10

---

9 The reason why the product of two symmetric matrices is not symmetric is the same as the reason why $\mathbf{R}$ from $\mathbf{QR}$ decomposition has zeros on the lower-diagonal.

Exercise 13-11.

Let's explore the scaling of eigenvectors. Start by creating a 4×4 matrix of random integers drawn between -14 and $+14$. Diagonalize the matrix and empirically confirm that $\mathbf{A} = \mathbf{V \Lambda V}^{-1}$. Confirm that the Euclidean norm of each eigenvector equals 1. Note that the square of a complex number is computed as that number times its complex conjugate (hint: use `np.conj()`).

Next, multiply the eigenvectors matrix by any nonzero scalar. I used π for no particularly good reason other than it was fun to type. Does this scalar affect the accuracy of the reconstructed matrix and/or the norms of the eigenvectors? Why or why not?

Finally, repeat this but use a symmetric matrix, and replace $\mathbf{V}^{-1}$ with $\mathbf{V}^{\mathrm{T}}$. Does this change the conclusion?

## CHAPTER 14

Singular Value Decomposition

The previous chapter was really dense! I tried my best to make it comprehensible and rigorous, without getting too bogged down in details that have less relevance for data science.

Fortunately, most of what you learned about eigendecomposition applies to the SVD. That means that this chapter will be easier and shorter.

The purpose of the SVD is to decompose a matrix into the product of three matrices, called the left singular vectors ($\mathbf{U}$), the singular values ($\mathbf{\Sigma}$), and the right singular vectors ($\mathbf{V}$):

$$\mathbf{A} = \mathbf{U}\mathbf{\Sigma}\mathbf{V}^{\mathrm{T}}$$

This decomposition should look similar to eigendecomposition. In fact, you can think of the SVD as a generalization of eigendecomposition to nonsquare matrices— or you can think of eigendecomposition as a special case of the SVD for square matrices.[1]

The singular values are comparable to eigenvalues, while the singular vectors matrices are comparable to eigenvectors (these two sets of quantities are the same under some circumstances that I will explain later).

The Big Picture of the SVD

I want to introduce you to the idea and interpretation of the matrices, and then later in the chapter I will explain how to compute the SVD.

---

1 The SVD is not the same as eigendecomposition for all square matrices; more on this later.

Figure 14-1 shows the overview of the SVD.

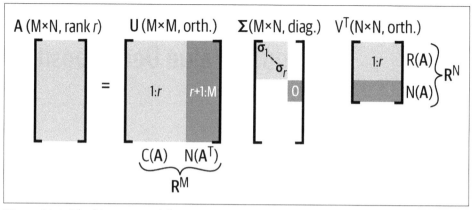

Figure 14-1. The big picture of the SVD

Many important features of the SVD are visible in this diagram; I will go into these features in more detail throughout this chapter, but to put them into a list:

- Both **U** and **V** are square matrices, even when **A** is nonsquare.

- The matrices of singular vectors **U** and **V** are orthogonal, meaning $\mathbf{U}^\mathrm{T}\mathbf{U} = \mathbf{I}$ and $\mathbf{V}^\mathrm{T}\mathbf{V} = \mathbf{I}$. As a reminder, this means that each column is orthogonal to each other column, and any subset of columns is orthogonal to any other (nonoverlapping) subset of columns.

- The first r columns of **U** provide orthogonal basis vectors for the column space of the matrix **A**, while the rest of the columns provide orthogonal basis vectors for the left-null space (unless $r = M$, in which case the matrix is full column-rank and the left-null space is empty).

- The first r rows of $\mathbf{V}^\mathrm{T}$ (which are the columns of **V**) provide orthogonal basis vectors for the row space, while the rest of the rows provide orthogonal basis vectors for the null space.

- The singular values matrix is a diagonal matrix of the same size as **A**. The singular values are always sorted from largest (top-left) to smallest (lower-right).

- All singular values are nonnegative and real-valued. They cannot be complex or negative, even if the matrix contains complex-valued numbers.

- The number of nonzero singular values equals the matrix rank.

Perhaps the most amazing thing about the SVD is that it reveals all four subspaces of the matrix: the column space and left-null space are spanned by the first r and last $M - r$ through M columns of **U**, while the row space and null space are spanned by the first r and last $N - r$ through N rows of $\mathbf{V}^T$. For a rectangular matrix, if $r = M$, then the left-null space is empty, and if $r = N$, then the null space is empty.

Singular Values and Matrix Rank

The rank of a matrix is defined as the number of nonzero singular values. The reason comes from the previous discussion that the column space and the row space of the matrix are defined as the left and right singular vectors that are scaled by their corresponding singular values to have some "volume" in the matrix space, whereas the left and right null spaces are defined as the left and right singular vectors that are scaled to zeros. Thus, the dimensionality of the column and row spaces are determined by the number of nonzero singular values.

In fact, we can peer into the NumPy function `np.linalg.matrix_rank` to see how Python computes matrix rank (I've edited the code slightly to focus on the key concepts):

```
S = svd(M,compute_uv=False) # return only singular values
tol = S.max() * max(M.shape[-2:]) * finfo(S.dtype).eps
return count_nonzero(S > tol)
```

The returned value is the number of singular values that exceed the value of `tol`. What is `tol`? That's a tolerance level that accounts for possible rounding errors. It is defined as the machine precision for this data type (`eps`), scaled by the largest singular value and the size of the matrix.

Thus, we yet again see the difference between "chalkboard math" and precision math implemented on computers: the rank of the matrix is not actually computed as the number of nonzero singular values, but instead as the number of singular values that are larger than some small number. There is a risk that small but truly nonzero singular values are ignored, but that outweighs the risk of incorrectly inflating the rank of the matrix when truly zero-valued singular values appear nonzero due to precision errors.

SVD in Python

The SVD is fairly straightforward to compute in Python:

```
U,s,Vt = np.linalg.svd(A)
```

There are two features of NumPy's `svd` function to keep in mind. First, the singular values are returned as a vector, not a matrix of the same size as $\mathbf{A}$. This means that you need some extra code to get the Σ matrix:

```
S = np.zeros(np.shape(A))
np.fill_diagonal(S,s)
```

You might initially think of using `np.diag(s)`, but that only produces the correct singular values matrix for a square matrix **A**. Therefore, I first create the correctly sized matrix of zeros, and then fill in the diagonal with the singular values.

The second feature is that NumPy returns the matrix $\mathbf{V}^{\mathrm{T}}$, not **V**. This may be confusing for readers coming from a MATLAB background, because the MATLAB `svd` function returns the matrix **V**. The hint is in the docstring, which describes matrix `vh`, where the `h` is for Hermitian, the name of a symmetric complex-valued matrix.

Figure 14-2 shows the outputs of the `svd` function (with the singular values converted into a matrix).

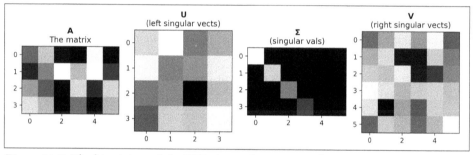

Figure 14-2. The big picture of the SVD shown for an example matrix

SVD and Rank-1 "Layers" of a Matrix

The first equation I showed in the previous chapter was the vector-scalar version of the eigenvalue equation ($\mathbf{Av} = \lambda\mathbf{v}$). I opened this chapter with the *matrix* SVD equation ($\mathbf{A} = \mathbf{U\Sigma V}^{\mathrm{T}}$); what does that equation look like for one vector? We can write it in two different ways that highlight different features of the SVD:

$$\mathbf{Av} = \mathbf{u}\sigma$$
$$\mathbf{u}^{\mathrm{T}}\mathbf{A} = \sigma\mathbf{v}^{\mathrm{T}}$$

Those equations are kind of similar to the eigenvalue equation except that there are two vectors instead of one. The interpretations are, therefore, slightly more nuanced: in general, those equations say that the effect of the matrix on one vector is the same as the effect of a scalar on a different vector.

Notice that the first equation means that **u** is in the column space of **A**, with **v** providing the weights for combining the columns. Same goes for the second equation, but **v** is in the row space of **A** with **u** providing the weights.

But that's not what I want to focus on in this section; I want to consider what happens when you multiply one left singular vector by one right singular vector. Because the singular vectors are paired with the same singular value, we need to multiply the ith left singular vector by the ith singular value by the ith right singular vector.

Note the orientations in this vector-vector multiplication: column on the left, row on the right (Figure 14-3). That means that the result will be an outer product of the same size as the original matrix. Furthermore, that outer product is a rank-1 matrix whose norm is determined by the singular value (because the singular vectors are unit-length):

$$\mathbf{u}_1 \sigma_1 \mathbf{v}_1^{\mathsf{T}} = \mathbf{A}_1$$

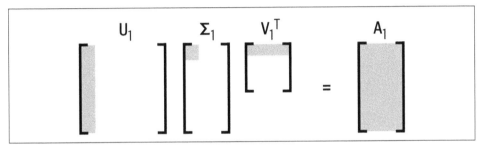

Figure 14-3. Outer product of singular vectors creates a matrix "layer"

The subscripted 1 in the equation indicates using the first singular vectors and first (largest) singular value. I call the result $\mathbf{A}_1$ because it's not the original matrix $\mathbf{A}$; instead, it's a rank-1 matrix of the same size as $\mathbf{A}$. And not just any rank-1 matrix—it is the most important "layer" of the matrix. It's the most important because it has the largest singular value (more on this point in a later section).

With this in mind, we can reconstruct the original matrix by summing all SVD "layers" associated with $\sigma > 0$:[2]

$$\mathbf{A} = \sum_{i=1}^{r} \mathbf{u}_i \sigma_i \mathbf{v}_i^{\mathsf{T}}$$

The point of showing this summation is that you don't necessarily need to use all r layers; instead, you can reconstruct some other matrix, let's call it $\widetilde{\mathbf{A}}$, which contains the first $k < r$ layers. This is called a *low-rank approximation* of matrix $\mathbf{A}$—in this case, a rank-k approximation.

---

2 There is no point in summing the zero-valued singular values, because that's just adding zeros matrices.

Low-rank approximations are used, for example, in data cleaning. The idea is that information associated with small singular values makes little contribution to the total variance of the dataset, and therefore might reflect noise that can be removed. More on this in the next chapter.

SVD from EIG

OK, at this point in the chapter you know the basics of understanding and interpreting the SVD matrices. I'm sure you are wondering what this magical formula is to produce the SVD. Perhaps it's so incredibly complicated that only Gauss could understand it? Or maybe it would take so long to explain that it doesn't fit into one chapter?

Wrong!

In fact, the SVD is really easy (conceptually; performing an SVD by hand is another matter). It simply comes from computing the eigendecomposition of the matrix times its transpose. The following equations show how to derive the singular values and the left singular vectors:

$$\mathbf{A}\mathbf{A}^{\mathrm{T}} = \left(\mathbf{U}\Sigma\mathbf{V}^{\mathrm{T}}\right)\left(\mathbf{U}\Sigma\mathbf{V}^{\mathrm{T}}\right)^{\mathrm{T}}$$
$$= \mathbf{U}\Sigma\mathbf{V}^{\mathrm{T}}\mathbf{V}\Sigma^{\mathrm{T}}\mathbf{U}^{\mathrm{T}}$$
$$= \mathbf{U}\Sigma^{2}\mathbf{U}^{\mathrm{T}}$$

In other words, the eigenvectors of $\mathbf{A}\mathbf{A}^{\mathrm{T}}$ are the left-singular vectors of $\mathbf{A}$, and the squared eigenvalues of $\mathbf{A}\mathbf{A}^{\mathrm{T}}$ are the singular values of $\mathbf{A}$.

This insight reveals three features of the SVD: (1) singular values are nonnegative because squared numbers cannot be negative, (2) singular values are real-valued because symmetric matrices have real-valued eigenvalues, and (3) singular vectors are orthogonal because the eigenvectors of a symmetric matrix are orthogonal.

The right-singular values come from premultiplying the matrix transpose:

$$\mathbf{A}^{\mathrm{T}}\mathbf{A} = \left(\mathbf{U}\Sigma\mathbf{V}^{\mathrm{T}}\right)^{\mathrm{T}}\left(\mathbf{U}\Sigma\mathbf{V}^{\mathrm{T}}\right)$$
$$= \mathbf{V}\Sigma^{\mathrm{T}}\mathbf{U}^{\mathrm{T}}\mathbf{U}\Sigma\mathbf{V}^{\mathrm{T}}$$
$$= \mathbf{V}\Sigma^{2}\mathbf{V}^{\mathrm{T}}$$

In fact, you can rearrange the SVD equation to solve for the right-singular vectors without having to compute the eigendecomposition of $\mathbf{A}^T\mathbf{A}$:

$$\mathbf{V}^T = \mathbf{\Sigma}^{-1}\mathbf{U}^T\mathbf{A}$$

Of course, there is a complementary equation for deriving $\mathbf{U}$ if you already know $\mathbf{V}$.

SVD of $\mathbf{A}^T\mathbf{A}$

Briefly, if a matrix can be expressed as $\mathbf{S} = \mathbf{A}^T\mathbf{A}$, then its left- and right-singular vectors are equal. In other words:

$$\mathbf{S} = \mathbf{U}\mathbf{\Sigma}\mathbf{V}^T = \mathbf{V}\mathbf{\Sigma}\mathbf{U}^T = \mathbf{U}\mathbf{\Sigma}\mathbf{U}^T = \mathbf{V}\mathbf{\Sigma}\mathbf{V}^T$$

The proof of this claim comes from writing out the SVD of $\mathbf{S}$ and $\mathbf{S}^T$, then considering the implication of $\mathbf{S} = \mathbf{S}^T$. I'm leaving this one for you to explore on your own! And I also encourage you to confirm it in Python using random symmetric matrices.

In fact, for a symmetric matrix, SVD is the same thing as eigendecomposition. This has implications for principal components analysis, because PCA can be performed using eigendecomposition of the data covariance matrix, the SVD of the covariance matrix, or the SVD of the data matrix.

Converting Singular Values to Variance, Explained

The sum of the singular values is the total amount of "variance" in the matrix. What does that mean? If you think of the information in the matrix as being contained in a bubble, then the sum of the singular values is like the volume of that bubble.

The reason why all the variance is contained in the singular values is that the singular vectors are normalized to unit magnitude, which means they provide no magnitude information (that is, $\| \mathbf{U}\mathbf{w} \| = \| \mathbf{w} \|$).[3] In other words, the singular vectors point, and the singular values say how far.

The "raw" singular values are in the numerical scale of the matrix. That means that if you multiply the data by a scalar, then the singular values will increase. And this in turn means that the singular values are difficult to interpret, and are basically impossible to compare across different datasets.

---

3 The proof of this statement is in Exercise 14-3.

For this reason, it is often useful to convert the singular values to percent total variance explained. The formula is simple; each singular value i is normalized as follows:

$$\widetilde{\sigma}_i = \frac{100\sigma_i}{\Sigma \sigma}$$

This normalization is common in principal components analysis, for example, to determine the number of components that account for 99% of the variance. That can be interpreted as an indicator of system complexity.

Importantly, this normalization does not affect the relative distances between singular values; it merely changes the numerical scale into one that is more readily intepretable.

Condition Number

I've hinted several times in this book that the condition number of a matrix is used to indicate the numerical stability of a matrix. Now that you know about singular values, you can better appreciate how to compute and interpret the condition number.

The condition number of a matrix is defined as the ratio of the largest to the smallest singular value. It's often given the letter κ (Greek letter *kappa*):

$$\kappa = \frac{\sigma_{max}}{\sigma_{min}}$$

The condition number is often used in statistics and machine learning to evaluate the stability of a matrix when computing its inverse and when using it to solve systems of equations (e.g., least squares). Of course, a noninvertible matrix has a condition number of NaN because $\sigma/0 = ?$.

But a numerically full-rank matrix with a large condition number can still be unstable. Though theoretically invertible, in practice the matrix inverse may be unreliable. Such matrices are called *ill-conditioned*. You might have seen that term in warning messages in Python, sometimes accompanied by phrases like "result is not guaranteed to be accurate."

What's the problem with an ill-conditioned matrix? As the condition number increases, the matrix tends toward being singular. Therefore, an ill-conditioned matrix is "almost singular" and its inverse becomes untrustworthy due to increased risk of numerical errors.

There are a few ways to think about the impact of an ill-conditioned matrix. One is as the decrease in the precision of a solution due to round-off errors. For example, a condition number on the order of 10^5 means that the solution (e.g., the matrix inverse or least squares problem) loses five significant digits (this would mean, for example, going from a precision of 10^{-16} to 10^{-11}).

A second interpretation, related to the previous, is as an amplification factor for noise. If you have a matrix with a condition number on the order of 10^4, then noise could impact the solution to a least squares problem by 10^4. That might sound like a lot, but it could be an insigificant amplification if your data has a precision of 10^{-16}.

Thirdly, the condition number indicates the sensitivity of a solution to perturbations in the data matrix. A well-conditioned matrix can be perturbed (more noise added) with minimal change in the solution. In contrast, adding a small amount of noise to an ill-conditioned matrix can lead to wildly different solutions.

What is the threshold for a matrix to be ill-conditioned? There is none. There is no magic number that separates a well-conditioned from an ill-conditioned matrix. Different algorithms will apply different thresholds that depend on the numerical values in the matrix.

This much is clear: take warning messages about ill-conditioned matrices seriously. They usually indicate that something is wrong and the results should not be trusted.

What to Do About Ill-Conditioned Matrices?

That's unfortunately not a question I can give a specific answer to. The right thing to do when you have an ill-conditioned matrix depends a lot on the matrix and on the problem you're trying to solve. To be clear: ill-conditioned matrices are not intrinsically bad, and you should never flippantly discard a matrix simply because of its condition number. An ill-conditioned matrix is only potentially problematic for certain operations, and therefore is only relevant for certain matrices, such as statistical design matrices or covariance matrices.

Treatment options for an ill-conditioned matrix include regularization, dimensionality reduction, and improving data quality or feature extraction.

SVD and the MP Pseudoinverse

The SVD of a matrix inverse is quite elegant. Assuming the matrix is square and invertible, we get:

$$\mathbf{A}^{-1} = \left(\mathbf{U}\boldsymbol{\Sigma}\mathbf{V}^{\mathrm{T}}\right)^{-1}$$
$$= \mathbf{V}\boldsymbol{\Sigma}^{-1}\mathbf{U}^{-1}$$
$$= \mathbf{V}\boldsymbol{\Sigma}^{-1}\mathbf{U}^{\mathrm{T}}$$

In other words, we only need to invert $\boldsymbol{\Sigma}$, because $\mathbf{U}^{-1} = \mathbf{U}^{\mathrm{T}}$. Furthermore, because $\boldsymbol{\Sigma}$ is a diagonal matrix, its inverse is obtained simply by inverting each diagonal element. On the other hand, this method is still subject to numerical instabilities, because tiny singular values that might reflect precision errors (e.g., 10^{-15}) become galatically large when inverted.

Now for the algorithm to compute the MP pseudoinverse. You've waited many chapters for this; I appreciate your patience.

The MP pseudoinverse is computed almost exactly as the full inverse shown in the previous example; the only modification is to invert the *nonzero* diagonal elements in $\boldsymbol{\Sigma}$ instead of trying to to invert all diagonal elements. (In practice, "nonzero" is implemented as above a threshold to account for precision errors.)

And that's it! That's how the pseudoinverse is computed. You can see that it's very simple and intuitive, but requires a considerable amount of background knowledge about linear algebra to understand.

Even better: because the SVD works on matrices of any size, the MP pseudoinverse can be applied to nonsquare matrices. In fact, the MP pseudoinverse of a tall matrix equals its left-inverse, and the MP pseudoinverse of a wide matrix equals its right-inverse. (Quick reminder that the pseudoinverse is indicated as $\mathbf{A}^+$, $\mathbf{A}^*$, or $\mathbf{A}^\dagger$.)

You'll gain more experience with the pseudoinverse by implementing it yourself in the exercises.

Summary

I hope you agree that after putting in the effort to learn about eigendecomposition, a little bit of extra effort goes a long way toward understanding the SVD. The SVD is arguably the most important decomposition in linear algebra, because it reveals rich and detailed information about the matrix. Here are the key points:

- SVD decomposes a matrix (of any size and rank) into the product of three matrices, termed *left singular vectors* $\mathbf{U}$, *singular values* $\boldsymbol{\Sigma}$, and *right singular vectors* $\mathbf{V}^{\mathrm{T}}$.

- The first r (where r is the matrix rank) left singular vectors provide an orthonormal basis set for the column space of the matrix, while the later singular vectors provide an orthonormal basis set for the left-null space.

- It's a similar story for the right singular vectors: the first r vectors provide an orthonormal basis set for the row space, while the later vectors provide an orthonormal basis set for the null space. Be mindful that the right singular vectors are actually the *rows* of $\mathbf{V}$, which are the *columns* of $\mathbf{V}^{\mathrm{T}}$.

- The number of nonzero singular values equals the rank of the matrix. In practice, it can be difficult to distinguish between very small nonzero singular values versus precision errors on zero-valued singular values. Programs like Python will use a tolerance threshold to make this distinction.

- The outer product of the kth left singular vector and the kth right singular vector, scalar multiplied by the kth singular value, produces a rank-1 matrix that can be interpreted as a "layer" of the matrix. Reconstructing a matrix based on layers has many applications, including denoising and data compression.

- Conceptually, the SVD can be obtained from the eigendecomposition of $\mathbf{AA}^T$.

- The super-duper important Moore-Penrose pseudoinverse is computed as $\mathbf{V\Sigma^+U}^T$, where $\mathbf{\Sigma}^+$ is obtained by inverting the nonzero singular values on the diagonal.

Code Exercises

Exercise 14-1.

You learned that for a symmetric matrix, the singular values and the eigenvalues are the same. How about the singular vectors and eigenvectors? Use Python to answer this question using a random 5×5 $\mathbf{A}^T\mathbf{A}$ matrix. Next, try it again using the additive method for creating a symmetric matrix $(\mathbf{A}^T + \mathbf{A})$. Pay attention to the signs of the eigenvalues of $\mathbf{A}^T + \mathbf{A}$.

Exercise 14-2.

Python can optionally return the "economy" SVD, which means that the singular vectors matrices are truncated at the smaller of M or N. Consult the docstring to figure out how to do this. Confirm with tall and wide matrices. Note that you would typically want to return the full matrices; economy SVD is mainly used for really large matrices and/or really limited computational power.

Exercise 14-3.

One of the important features of an orthogonal matrix (such as the left and right singular vectors matrices) is that they rotate, but do not scale, a vector. This means that the magnitude of a vector is preserved after multiplication by an orthogonal matrix. Prove that $\| \mathbf{Uw} \| = \| \mathbf{w} \|$. Then demonstrate this empirically in Python by using a singular vectors matrix from the SVD of a random matrix and a random vector.

Exercise 14-4.

Create a random tall matrix with a specified condition number. Do this by creating two random square matrices to be **U** and **V**, and a rectangular **Σ**. Confirm that the empirical condition number of $\mathbf{U\Sigma V}^{\mathrm{T}}$ is the same as the number you specified. Visualize your results in a figure like Figure 14-4. (I used a condition number of 42.[4])

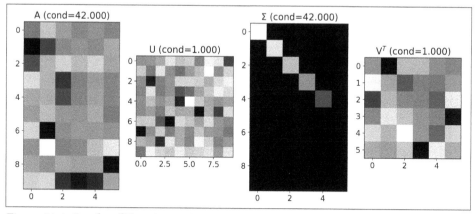

Figure 14-4. Results of Exercise 14-3

Exercise 14-5.

Your goal here is simple: write code to reproduce Figure 14-5. What does this figure show? Panel A shows a 30×40 random matrix that I created by smoothing random numbers (implemented as the 2D convolution between a 2D Gaussian and random numbers; if you are not familiar with image processing and filtering, then please feel free to copy the code to create this matrix from my code solution). The rest of panel A shows the SVD matrices. It's interesting to note that the earlier singular vectors (associated with the larger singular values) are smoother while the later ones are more rugged; this comes from the spatial filtering.

Panel B shows a "scree plot," which is the singular values normalized to percent variance explained. Notice that the first few components account for most of the variance in the image, while the later components each account for relatively little variance. Confirm that the sum over all normalized singular values is 100. Panel C shows the first four "layers"—rank-1 matrices defined as $\mathbf{u}_i\sigma_i\mathbf{v}_i^{\mathrm{T}}$—on the top row and the cumulative sum of those layers on the bottom row. You can see that each layer adds more information to the matrix; the lower-right image (titled "L 0:3") is a rank-4 matrix and yet appears visually very similar to the original rank-30 matrix in panel A.

---

4 Yes, that's another reference to *Hitchhiker's Guide to the Galaxy*.

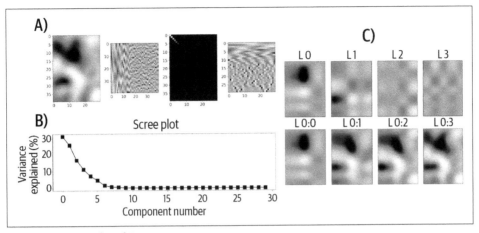

Figure 14-5. Results of Exercise 14-5

Exercise 14-6.

Implement the MP pseudoinverse based on the description in this chapter. You'll need to define a tolerance to ignore tiny-but-nonzero singular values. Please don't look up NumPy's implementation—and don't check back to earlier code in this chapter—but instead use your knowledge of linear algebra to come up with your own tolerance.

Test your code on a 5×5 rank-3 matrix. Compare your result against the output of NumPy's `pinv` function. Finally, inspect the source code for `np.linalg.pinv` to make sure you understand the implementation.

Exercise 14-7.

Demonstrate that the MP pseudoinverse equals the left-inverse for a full column-rank matrix by computing the explicit left-inverse of a tall full matrix $\left(\left(\mathbf{A}^\mathsf{T}\mathbf{A}\right)^{-1}\mathbf{A}^\mathsf{T}\right)$ and the pseudoinverse of $\mathbf{A}$. Repeat for the right inverse with a wide full row-rank matrix.

Example 14-8.

Consider the eigenvalue equation $\mathbf{Av} = \lambda\mathbf{v}$. Now that you know about the pseudoinverse, you can play around with that equation a bit. In particular, use the 2×2 matrix used at the outset of Chapter 13 to compute $\mathbf{v}^+$ and confirm that $\mathbf{vv}^+ = 1$. Next, confirm the following identities:

$$\mathbf{v}^+\mathbf{Av} = \lambda\mathbf{v}^+\mathbf{v}$$
$$\mathbf{Avv}^+ = \lambda\mathbf{vv}^+$$

## CHAPTER 15

Eigendecomposition and SVD Applications

Eigendecomposition and SVD are gems that linear algebra has bestowed upon modern human civilization. Their importance in modern applied mathematics cannot be understated, and their applications are uncountable and spread across myriad disciplines.

In this chapter, I will highlight three applications that you are likely to come across in data science and related fields. My main goal is to show you that seemingly complicated data science and machine learning techniques are actually quite sensible and easily understood, once you've learned the linear algebra topics in this book.

PCA Using Eigendecomposition and SVD

The purpose of PCA is to find a set of basis vectors for a dataset that point in the direction that maximizes covariation across the variables.

Imagine that an N-D dataset exists in an N-D space, with each data point being a coordinate in that space. This is sensible when you think about storing the data in a matrix with N observations (each row is an observation) of M features (each column is a feature, also called variable or measurement); the data live in $\mathbb{R}^M$ and comprise N vectors or coordinates.

An example in 2D is shown in Figure 15-1. The left-side panel shows the data in its original data space, in which each variable provides a basis vector for the data. Clearly the two variables (the x- and y-axes) are related to each other, and clearly there is a direction in the data that captures that relation better than either of the feature basis vectors.

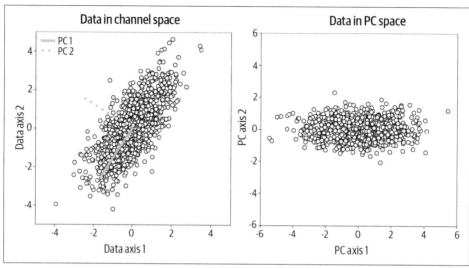

Figure 15-1. Graphical overview of PCA in 2D

The goal of PCA is to find a new set of basis vectors such that the linear relationships across the variables are maximally aligned with the basis vectors—that's what the right-side panel of Figure 15-1 shows. Importantly, PCA has the constraint that the new basis vectors are orthogonal rotations of the original basis vectors. In the exercises, you will see the implications of this constraint.

In the next section, I will introduce the math and procedures for computing PCA; in the exercises, you will have the opportunity to implement PCA using eigendecomposition and SVD, and compare your results against Python's implementation of PCA.

The Math of PCA

PCA combines the statistical concept of variance with the linear algebra concept of linear weighted combination. Variance, as you know, is a measure of the dispersion of a dataset around its average value. PCA makes the assumption that variance is good, and directions in the data space that have more variance are more important (a.k.a. "variance = relevance").

But in PCA, we're not just interested in the variance *within* one variable; instead, we want to find the linear weighted combination *across* all variables that maximizes variance of that component (a *component* is a linear weighted combination of variables).

Let's write this down in math. Matrix $\mathbf{X}$ is our data matrix (a tall full column-rank matrix of observations by features), and $\mathbf{w}$ is the vector of weights. Our goal in PCA is to find the set of weights in $\mathbf{w}$ such that $\mathbf{Xw}$ has maximal variance. Variance is a scalar, so we can write that down as:

$$\lambda = \| \mathbf{Xw} \|^2$$

The squared vector norm is actually the same thing as variance when the data is mean-centered (that is, each data variable has a mean of zero);[1] I've omitted a scaling factor of $1/(N-1)$, because it does not affect the solution to our optimization goal.

The problem with that equation is you can simply set $\mathbf{w}$ to be HUGE numbers; the bigger the weights, the larger the variance. The solution is to scale the norm of the weighted combination of data variables by the norm of the weights:

$$\lambda = \frac{\| \mathbf{Xw} \|^2}{\| \mathbf{w} \|^2}$$

Now we have a ratio of two vector norms. We can expand those norms into dot products to gain some insight into the equation:

$$\lambda = \frac{\mathbf{w}^T\mathbf{X}^T\mathbf{Xw}}{\mathbf{w}^T\mathbf{w}}$$

$$\mathbf{C} = \mathbf{X}^T\mathbf{X}$$

$$\lambda = \frac{\mathbf{w}^T\mathbf{Cw}}{\mathbf{w}^T\mathbf{w}}$$

We've now discovered that the solution to PCA is the same as the solution to finding the directional vector that maximizes the *normalized* quadratic form (the vector norm is the normalization term) of the data covariance matrix.

That's all fine, but how do we actually find the elements in vector $\mathbf{w}$ that maximize λ?

The linear algebra approach here is to consider not just a single vector solution but an entire set of solutions. Thus, we rewrite the equation using matrix $\mathbf{W}$ instead of vector $\mathbf{w}$. That would give a matrix in the denominator, which is not a valid operation in linear algebra; therefore, we multiply by the inverse:

$$\mathbf{\Lambda} = \left(\mathbf{W}^T\mathbf{W}\right)^{-1}\mathbf{W}^T\mathbf{CW}$$

---

1 The online code demonstrates this point.

From here, we apply some algebra and see what happens:

$$\Lambda = \left(\mathbf{W}^{\mathrm{T}}\mathbf{W}\right)^{-1}\mathbf{W}^{\mathrm{T}}\mathbf{C}\mathbf{W}$$
$$\Lambda = \mathbf{W}^{-1}\mathbf{W}^{-\mathrm{T}}\mathbf{W}^{\mathrm{T}}\mathbf{C}\mathbf{W}$$
$$\Lambda = \mathbf{W}^{-1}\mathbf{C}\mathbf{W}$$
$$\mathbf{W}\Lambda = \mathbf{C}\mathbf{W}$$

Remarkably, we've discovered that the solution to PCA is to perform an eigendecomposition on the data covariance matrix. The eigenvectors are the weights for the data variables, and their corresponding eigenvalues are the variances of the data along each direction (each column of $\mathbf{W}$).

Because covariance matrices are symmetric, their eigenvectors—and therefore principal components—are orthogonal. This has important implications for the appropriateness of PCA for data analysis, which you will discover in the exercises.

PCA Proof

I'm going to write out the proof that eigendecomposition solves the PCA optimization goal. If you are not familiar with calculus and Lagrange multipliers, then please feel free to skip this box; I include it here in the interest of completeness, not because you need to understand it to solve the exercises or use PCA in practice.

Our goal is to maximize $\mathbf{w}^{\mathrm{T}}\mathbf{C}\mathbf{w}$ subject to the constraint that $\mathbf{w}^{\mathrm{T}}\mathbf{w} = 1$. We can express this optimization using a Lagrange multiplier:

$$L(\mathbf{w}, \lambda) = \mathbf{w}^{\mathrm{T}}\mathbf{C}\mathbf{w} - \lambda\left(\mathbf{w}^{\mathrm{T}}\mathbf{w} - 1\right)$$
$$0 = \frac{d}{d\mathbf{w}}\left(\mathbf{w}^{\mathrm{T}}\mathbf{C}\mathbf{w} - \lambda\left(\mathbf{w}^{\mathrm{T}}\mathbf{w} - 1\right)\right)$$
$$0 = \mathbf{C}\mathbf{w} - \lambda\mathbf{w}$$
$$\mathbf{C}\mathbf{w} = \lambda\mathbf{w}$$

Briefly, the idea is that we use the Lagrange multiplier to balance the optimization with the constraint, take the derivative with respect to the weights vector, set the derivative to zero, differentiate with respect to $\mathbf{w}$, and discover that $\mathbf{w}$ is an eigenvector of the covariance matrix.

The Steps to Perform a PCA

With the math out of the way, here are the steps to implement a PCA:[2]

1. Compute the covariance matrix of the data. The resulting covariance matrix will be features-by-features. Each feature in the data must be mean-centered prior to computing covariance.

2. Take the eigendecomposition of that covariance matrix.

3. Sort the eigenvalues descending by magnitude, and sort the eigenvectors accordingly. Eigenvalues of the PCA are sometimes called *latent factor scores*.

4. Compute the "component scores" as the weighted combination of all data features, where the eigenvector provides the weights. The eigenvector associated with the largest eigenvalue is the "most important" component, meaning the one with the largest variance.

5. Convert the eigenvalues to percent variance explained to facilitate interpretation.

PCA via SVD

PCA can equivalently be performed via eigendecomposition as previously described or via SVD. There are two ways to perform a PCA using SVD:

- Take the SVD of the covariance matrix. The procedure is identical to that previously described, because SVD and eigendecomposition are the same decomposition for covariance matrices.

- Take the SVD of the data matrix directly. In this case, the right singular vectors (matrix $\mathbf{V}$) are equivalent to the eigenvectors of the covariance matrix (it would be the left singular vectors if the data matrix is stored as features-by-observations). The data must be mean-centered before computing the SVD. The square root of the singular values is equivalent to the eigenvalues of the covariance matrix.

Should you use eigendecomposition or SVD to perform a PCA? You might think that SVD is easier because it does not require the covariance matrix. That's true for relatively small and clean datasets. But larger or more complicated datasets may require data selection or may be too memory intensive to take the SVD of the entire data matrix. In these cases, computing the covariance matrix first can increase analysis flexibility. But the choice of eigendecomposition versus SVD is often a matter of personal preference.

---

2 In Exercise 15-3, you will also learn how to implement PCA using the Python scikit-learn library.

Linear Discriminant Analysis

Linear discriminant analysis (LDA) is a multivariate classification technique that is often used in machine learning and statistics. It was initially developed by Ronald Fisher,[3] who is often considered the "grandfather" of statistics for his numerous and important contributions to the mathematical foundations of statistics.

The goal of LDA is to find a direction in the data space that maximally separates categories of data. An example problem dataset is shown in graph A in Figure 15-2. It is visually obvious that the two categories are separable, but they are not separable on either of the data axes alone—that is clear from visual inspection of the marginal distributions.

Enter LDA. LDA will find basis vectors in the data space that maximally separate the two categories. Graph B in Figure 15-2 shows the same data but in the LDA space. Now the classification is simple—observations with negative values on axis-1 are labeled category "0" and any observations with positive values on axis 1 are labeled category "1." The data is completely inseparable on axis 2, indicating that one dimension is sufficient for accurate categorization in this dataset.

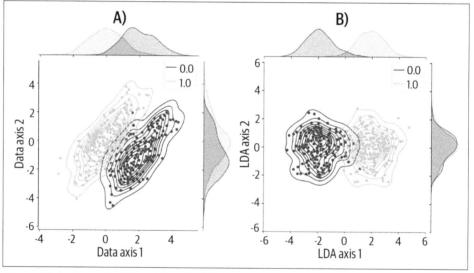

Figure 15-2. Example 2D problem for LDA

---

3 Indeed, linear discriminant analysis is also called Fisher's discriminant analysis.

Sounds great, right? But how does such a marvel of mathematics work? It's actually fairly straightforward and based on generalized eigendecomposition, which you learned about toward the end of Chapter 13.

Let me begin with the objective function: our goal is to find a set of weights such that the weighted combination of variables maximally separates the categories. That objective function can be written similarly as with the PCA objective function:

$$\lambda = \frac{\| \mathbf{X}_B\mathbf{w} \|^2}{\| \mathbf{X}_W\mathbf{w} \|^2}$$

In English: we want to find a set of feature weights $\mathbf{w}$ that maximizes the *ratio* of the variance of data feature $\mathbf{X}_B$, to the variance of data feature $\mathbf{X}_W$. Notice that the same weights are applied to all data observations. (I'll write more about data features B and W after discussing the math.)

The linear algebra solution comes from following a similar argument as described in the PCA section. First, expand $\| \mathbf{X}_B\mathbf{w} \|^2$ to $\mathbf{w}^{\mathsf{T}}\mathbf{X}_B^{\mathsf{T}}\mathbf{X}_B\mathbf{w}$ and express this as $\mathbf{w}^{\mathsf{T}}\mathbf{C}_B\mathbf{w}$; second, consider a set of solutions instead of one solution; third, replace the division with multiplication of the inverse; and finally, do some algebra and see what happens:

$$\mathbf{\Lambda} = \left(\mathbf{W}^{\mathsf{T}}\mathbf{C}_W\mathbf{W}\right)^{-1}\mathbf{W}^{\mathsf{T}}\mathbf{C}_B\mathbf{W}$$

$$\mathbf{\Lambda} = \mathbf{W}^{-1}\mathbf{C}_W^{-1}\mathbf{W}^{-\mathsf{T}}\mathbf{W}^{\mathsf{T}}\mathbf{C}_B\mathbf{W}$$

$$\mathbf{\Lambda} = \mathbf{W}^{-1}\mathbf{C}_W^{-1}\mathbf{C}_B\mathbf{W}$$

$$\mathbf{W}\mathbf{\Lambda} = \mathbf{C}_W^{-1}\mathbf{C}_B\mathbf{W}$$

$$\mathbf{C}_W\mathbf{W}\mathbf{\Lambda} = \mathbf{C}_B\mathbf{W}$$

In other words, the solution to LDA comes from a generalized eigendecomposition on two covariance matrices. The eigenvectors are the weights, and the generalized eigenvalues are the variance ratios of each component.[4]

With the math out of the way, which data features are used to construct $\mathbf{X}_B$ and $\mathbf{X}_W$? Well, there are different ways of implementing that formula, depending on the nature of the problem and the specific goal of the analysis. But in a typical LDA model, the $\mathbf{X}_B$ comes from the between-category covariance while the $\mathbf{X}_W$ comes from the within-category covariance.

---

4 I won't go through the calculus-laden proof, but it's just a minor variant of the proof given in the PCA section.

The within-category covariance is simply the average of the covariances of the data samples within each class. The between-category covariance comes from creating a new data matrix comprising the feature averages within each class. I will walk you through the procedure in the exercises. If you are familiar with statistics, then you'll recognize this formulation as analogous to the ratio of between-group to within-group sum of squared errors in ANOVA models.

Two final comments: The eigenvectors of a generalized eigendecomposition are not constrained to be orthogonal. That's because $\mathbf{C}_W^{-1}\mathbf{C}_B$ is generally not a symmetric matrix even though the two covariance matrices are themselves symmetric. Nonsymmetric matrices do not have the orthogonal-eigenvector constraint. You'll see this in the exercises.

Finally, LDA will always find a *linear* solution (duh, that's right in the name *LDA*), even if the data is *not* linearly separable. Nonlinear separation would require a transformation of the data or the use of a nonlinear categorization method like artificial neural networks. LDA will still work in the sense of producing a result; it's up to you as the data scientist to determine whether that result is appropriate and interpretable for a given problem.

Low-Rank Approximations via SVD

I explained the concept of low-rank approximations in the previous chapter (e.g., Exercise 14-5). The idea is to take the SVD of a data matrix or image, and then reconstruct that data matrix using some subset of SVD components.

You can achieve this by setting selected σs to equal zero or by creating new SVD matrices that are rectangular, with the to-be-rejected vectors and singular values removed. This second approach is preferred because it reduces the sizes of the data to be stored, as you will see in the exercises. In this way, the SVD can be used to compress data down to a smaller size.

Do Computers Use SVD to Compress Images?

The short answer is no, they don't.

Algorithms underlying common image compression formats like JPG involve blockwise compression that incorporates principles of human perception (including, for example, how we perceive contrast and spatial frequencies), allowing them to achieve better results with less computational power, compared to taking one SVD of the entire image.

Nonetheless, the principle is the same as with SVD-based compression: identify a small number of basis vectors that preserves the important features of an image such that the low-rank reconstruction is an accurate approximation of the full-resolution original.

That said, SVD for data compression is commonly used in other areas of science, including biomedical imaging.

SVD for Denoising

Denoising via SVD is simply an application of low-rank approximation. The only difference is that SVD components are selected for exclusion based on them representing noise as opposed to making small contributions to the data matrix.

The to-be-removed components might be layers associated with the smallest singular values—that would be the case for low-amplitude noise associated with small equipment imperfections. But larger sources of noise that have a stronger impact on the data might have larger singular values. These noise components can be identified by an algorithm based on their characteristics or by visual inspection. In the exercises, you will see an example of using SVD to separate a source of noise that was added to an image.

Summary

You've made it to the end of the book (except for the exercises below)! Congrats! Take a moment to be proud of yourself and your commitment to learning and investing in your brain (it is, after all, your most precious resource). I am proud of you, and if we would meet in person, I'd give you a high five, fist bump, elbow tap, or whatever is socially/medically appropriate at the time.

I hope you feel that this chapter helped you see the incredible importance of eigendecomposition and singular value decomposition to applications in statistics and machine learning. Here's a summary of the key points that I am contractually obligated to include:

- The goal of PCA is to find a set of weights such that the linear weighted combination of data features has maximal variance. That goal reflects the assumption underlying PCA, which is that "variance equals relevance."

- PCA is implemented as the eigendecomposition of a data covariance matrix. The eigenvectors are the feature weightings, and the eigenvalues can be scaled to encode percent variance accounted for by each component (a *component* is the linear weighted combination).

- PCA can be equivalently implemented using the SVD of the covariance matrix or the data matrix.

- Linear discriminant analysis (LDA) is used for linear categorization of multivariable data. It can be seen as an extension of PCA: whereas PCA maximizes variance, LDA maximizes the ratio of variances between two data features.

- LDA is implemented as a generalized eigendecomposition on two covariance matrices that are formed from two different data features. The two data features are often the between-class covariance (to maximize) and the within-class covariance (to minimize).

- Low-rank approximations involve reproducing a matrix from a subset of singular vectors/values and are used for data compression and denoising.

- For data compression, the components associated with the smallest singular values are removed; for data denoising, components that capture noise or artifacts are removed (their corresponding singular values could be small or large).

Exercises

PCA

I love Turkish coffee. It's made with very finely ground beans and no filter. The whole ritual of making it and drinking it is wonderful. And if you drink it with a Turk, perhaps you can have your fortune read.

This exercise is not about Turkish coffee, but it is about doing a PCA on a dataset[5] that contains time series data from the Istanbul stock exchange, along with stock exchange data from several other stock indices in different countries. We could use this dataset to ask, for example, whether the international stock exchanges are driven by one common factor, or whether different countries have independent financial markets.

Exercise 15-1.

Before performing a PCA, import and inspect the data. I made several plots of the data shown in Figure 15-3; you are welcome to reproduce these plots and/or use different methods to explore the data.

---

5 Data citation: Akbilgic, Oguz. (2013). Istanbul Stock Exchange. UCI Machine Learning Repository. Data source website: *https://archive-beta.ics.uci.edu/ml/datasets/istanbul+stock+exchange*.

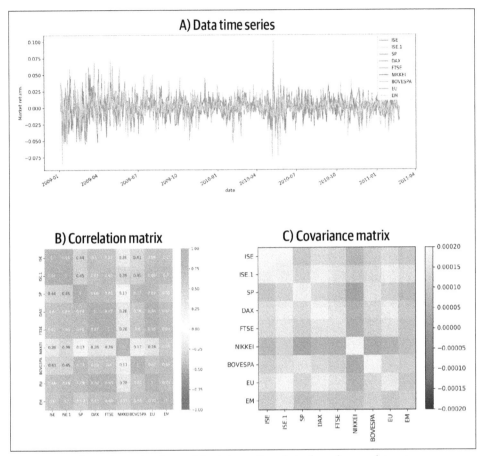

Figure 15-3. Some investigations of the international stock exchange dataset

Now for the PCA. Implement the PCA using the five steps outlined earlier in this chapter. Visualize the results as in Figure 15-4. Use code to demonstrate several features of PCA:

1. The variance of the component time series (using `np.var`) equals the eigenvalue associated with that component. You can see the results for the first two components here:

   ```
   Variance of first two components:
   [0.0013006  0.00028585]

   First two eigenvalues:
   [0.0013006  0.00028585]
   ```

2. The correlation between principal components (that is, the weighted combinations of the stock exchanges) 1 and 2 is zero, i.e., orthogonal.

3. Visualize the eigenvector weights for the first two components. The weights show how much each variable contributes to the component.

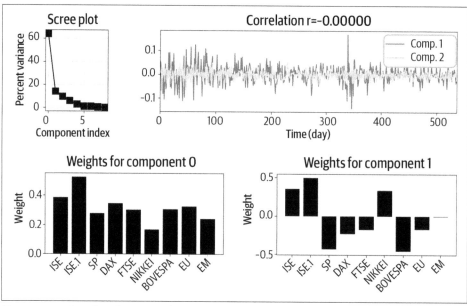

Figure 15-4. Results of PCA on the Instanbul Stock Exchange dataset

Discussion: The scree plot strongly suggests that the international stock exchanges are driven by a common factor of the global economy: there is one large component that accounts for around 64% of the variance in the data, while the other components each account for less than 15% of the variance (in a purely random dataset we would expect each component to account for $100/9 = 11\%$ of the variance, plus/minus noise).

A rigorous evaluation of the statistical significance of these components is outside the scope of this book, but based on visual inspection of the scree plot, we are not really justified to interpret the components after the first one; it appears that most of the variance in this dataset fits neatly into one dimension.

From the perspective of dimensionality reduction, we could reduce the entire dataset to the component associated with the largest eigenvalue (this is often called the *top component*), thereby representing this 9D dataset using a 1D vector. Of course, we lose information—36% of the information in the dataset is removed if we focus only on the top component—but hopefully, the important features of the signal are in the top component while the less important features, including random noise, are ignored.

Exercise 15-2.

Reproduce the results using (1) the SVD of the data covariance matrix and (2) the SVD of the data matrix itself. Remember that the eigenvalues of $\mathbf{X}^T\mathbf{X}$ are the squared singular values of $\mathbf{X}$; furthermore, the scaling factor on the covariance matrix must be applied to the singular values to find equivaluence.

Exercise 15-3.

Compare your "manual" PCA with the output of Python's PCA routine. You'll have to do some online research to figure out how to run a PCA in Python (this is one of the most important skills in Python programming!), but I'll give you a hint: it's in the sklearn.decomposition library.

sklearn or Manual Implementation of PCA?

Should you compute PCA by writing code to compute and eigendecompose the covariance matrix or use sklearn's implementation? There is always a trade-off between using your own code to maximize customization versus using prepackaged code to maximize ease. One of the myriad and amazing benefits of understanding the math behind data science analyses is that you can custom tailor analyses to suit your needs. In my own research, I find that implementing PCA on my own gives me more freedom and flexibility.

Exercise 15-4.

Now you will perform a PCA on simulated data, which will highlight one of the potential limitations of PCA. The goal is to create a dataset comprising two "streams" of data and plot the principal components on top, like in Figure 15-5.

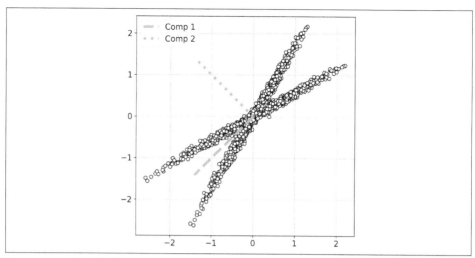

Figure 15-5. Results from Exercise 15-4

Here's how to create the data:

1. Create a 1,000 × 2 matrix of random numbers drawn from a normal (Gaussian) distribution in which the second column is scaled down by .05.

2. Create a 2 × 2 pure rotation matrix (see Chapter 7).

3. Stack two copies of the data vertically: once with the data rotated by $\theta = -\pi/6$, and once with the data rotated by $\theta = -\pi/3$. The resulting data matrix will be of size 2,000 × 2.

Use SVD to implement the PCA. I scaled the singular vectors by a factor of 2 for visual inspection.

Discussion: PCA is excellent for reducing dimensionality of a high-dimensional dataset. This can facilitate data compression, data cleaning, and numerical stability issues (e.g., imagine that a 200-dimensional dataset with a condition number of 10^{10} is reduced to the largest 100 dimensions with a condition number of 10^5). But the dimensions themselves may be poor choices for feature extraction, because of the orthogonality constraint. Indeed, the principal directions of variance in Figure 15-5 are correct in a mathematical sense, but I'm sure you have the feeling that those are not the best basis vectors to capture the features of the data.

Linear Discriminant Analyses

Exercise 15-5.

You are going to perform an LDA on simulated 2D data. Simulated data is advantageous because you can manipulate the effects sizes, amount and nature of noise, number of categories, and so on.

The data you will create was shown in Figure 15-2. Create two sets of normally distributed random numbers, each of size 200×2, with the second dimension being added onto the first (this imposes a correlation between the variables). Then add an xy offset of $[2 -1]$ to the first set of numbers. It will be convenient to create a 400×2 matrix that contains both data classes, as well as a 400-element vector of class labels (I used 0s for the first class and 1s for the second class).

Use `sns.jointplot` and `plot_joint` to reproduce graph A in Figure 15-2.

Exercise 15-6.

Now for the LDA. Write code in NumPy and/or SciPy instead of using a built-in library such as sklearn (we'll get to that later).

The within-class covariance matrix $\mathbf{C}_W$ is created by computing the covariance of each class separately and then averaging those covariance matrices. The between-class covariance matrix $\mathbf{C}_B$ is created by computing the means of each data feature (in this case, the xy-coordinates) within each class, concatenating those feature-mean vectors for all classes (that will create a 2×2 matrix for two features and two classes), and then computing the covariance matrix of that concatenated matrix.

Remember from Chapter 13 that generalized eigendecomposition is implemented using SciPy's `eigh` function.

The data projected into the LDA space is computed as $\widetilde{\mathbf{X}}\mathbf{V}$, where $\widetilde{\mathbf{X}}$ contains the concatenated data from all classes, mean-centered per feature, and $\mathbf{V}$ is the matrix of eigenvectors.

Compute the classification accuracy, which is simply whether each data sample has a negative ("class 0") or positive ("class 1") projection onto the first LDA component. Graph C in Figure 15-6 shows the predicted class label for each data sample.

Finally, show the results as shown in Figure 15-6.

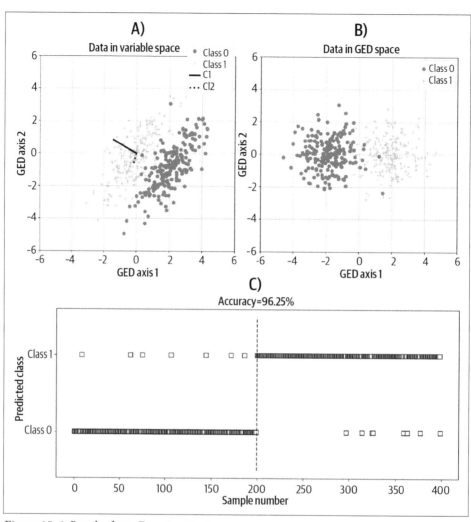

Figure 15-6. Results from Exercise 15-6

Exercise 15-7.

I claimed in Chapter 13 that for a generalized eigendecomposition, the matrix of eigenvectors $\mathbf{V}$ is not orthogonal, but it is orthogonal in the space of the "denominator" matrix. Your goal here is to demonstrate that empirically.

Compute and inspect the results of $\mathbf{V}^T\mathbf{V}$ and $\mathbf{V}^T\mathbf{C}_W\mathbf{V}$. Ignoring tiny precision errors, which one produces the identity matrix?

Exercise 15-8.

Now to reproduce our results using Python's sklearn library. Use the `LinearDiscri minantAnalysis` function in `sklearn.discriminant_analysis`. Produce a plot like Figure 15-7 and confirm that the overall prediction accuracy matches results from your "manual" LDA analysis in the previous exercise. This function allows for several different solvers; use the `eigen` solver to match the previous exercise, and also to complete the following exercise.

Plot the predicted labels from your "manual" LDA on top; you should find that the predicted labels are the same from both approaches.

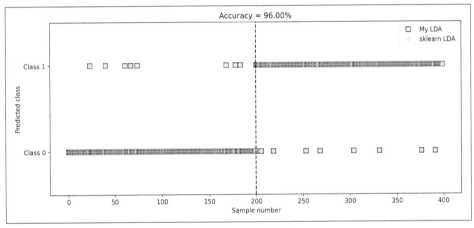

Figure 15-7. Results from Exercise 15-8

Exercise 15-9.

Let's use sklearn to explore the effects of shrinkage regularization. As I wrote in Chapters 12 and 13, it is trivial that shrinkage will reduce performance on training data; the important question is whether the regularization improves prediction accuracy on unseen data (sometimes called a *validation set* or *test set*). Therefore, you should write code to implement train/test splits. I did this by randomly permuting sample indices between 0 and 399, training on the first 350, and then testing on the final 50. Because this is a small number of samples to average, I repeated this random selection 50 times and took the average accuracy to be the accuracy per shrinkage amount in Figure 15-8.

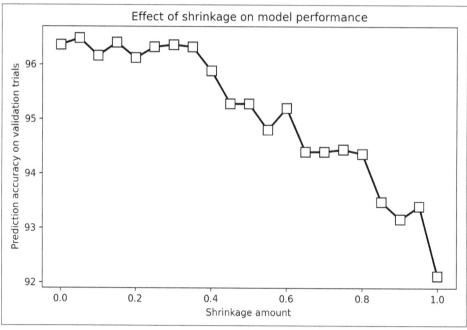

Figure 15-8. Results from Exercise 15-9

Discussion: Shrinkage generally had a negative impact on validation performance. Although it looks like the performance improved with some shrinkage, repeating the code multiple times showed that these were just some random fluctuations. A deeper dive into regularization is more appropriate for a dedicated machine learning book, but I wanted to highlight here that many "tricks" that have been developed in machine learning are not necessarily advantageous in all cases.

SVD for Low-Rank Approximations

Exercise 15-10.

Igor Stravinsky was one of the greatest music composers of all time (IMHO)—certainly one of the most influential of the 20th century. He also made many thought-provoking statements on the nature of art, media, and criticism, including one of my favorite quotes: "The more art is limited, the more it is free." There is a famous and captivating drawing of Stravinsky by none other than the great Pablo Picasso. An image of this drawing is available on Wikipedia (*https://oreil.ly/BtSZv*), and we are going to work with this picture in the next several exercises. Like with other images we've worked with in this book, it is natively a 3D matrix ($640 \times 430 \times 3$), but we will convert it to grayscale (2D) for convenience.

The purpose of this exercise is to repeat Exercise 14-5, in which you re-created a close approximation to a smooth-noise image based on four "layers" from the SVD (please look back at that exercise to refresh your memory). Produce a figure like Figure 15-9 using the Stravinsky image. Here is the main question: does reconstructing the image using the first four components give a good result like it did in the previous chapter?

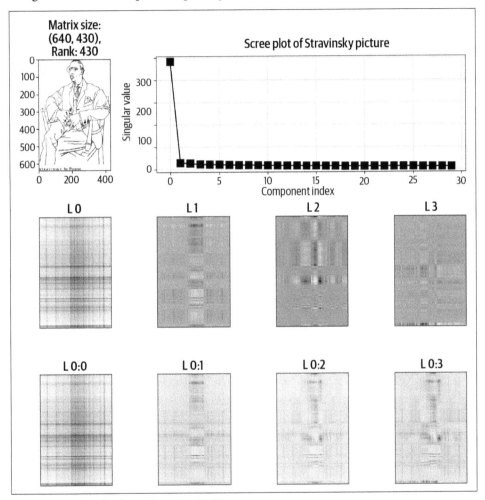

Figure 15-9. Results from Exercise 15-10

Exercise 15-11.

Well, the answer to the question at the end of the previous exercise is a resounding "No!" The rank-4 approximation is terrible! It looks nothing like the original image. The goal of this exercise is to reconstruct the image using more layers so that the low-rank approximation is reasonably accurate—and then compute the amount of compression obtained.

Start by producing Figure 15-10, which shows the original image, the reconstructed image, and the error map, which is the squared difference between the original and the approximation. For this figure, I chose $k = 80$ components, but I encourage you to explore different values (that is, different rank approximations).

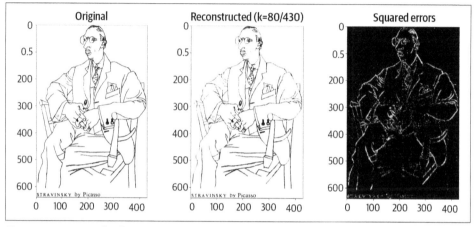

Figure 15-10. Results from Exercise 15-11

Next, compute the compression ratio, which is the percentage of the number of bytes used by the low-rank approximation versus the number of bytes used by the original image. My results for $k = 80$ are shown here.[6] Keep in mind that with low-rank approximations, you don't need to store the full image or the full SVD matrices!

```
       Original is 2.10 mb
  Reconstruction is 2.10 mb
  Recon vectors are 0.65 mb (using k=80 comps.)

  Compression of 31.13%
```

Exercise 15-12.

Why did I choose $k = 80$ and not, e.g., 70 or 103? It was quite arbitrary, to be honest. The goal of this exercise is to see whether it's possible to use the error map to determine an appropriate rank parameter.

In a for loop over reconstruction ranks between 1 and the number of singular values, create the low-rank approximation and compute the Frobenius distance between the

---

6 There is some ambiguity about whether one megabyte is $1,000^2$ or $1,024^2$ bytes; I used the latter, but it doesn't affect the compression ratio.

original and *k*-rank approximation. Then make a plot of the error as a function of rank, as in Figure 15-11. The error certainly decreases with increasing rank, but there is no clear rank that seems best. Sometimes in optimization algorithms, the derivative of the error function is more informative; give that a try!

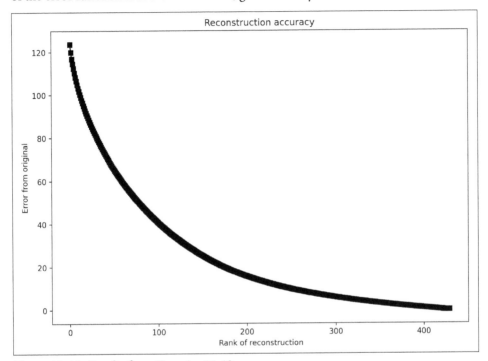

Figure 15-11. Results from Exercise 15-12

Final thought for this exercise: the reconstruction error for $k = 430$ (i.e., the full SVD) should be exactly 0. Is it? Obviously the answer is no; otherwise, I wouldn't have written the question. But you should confirm this yourself. This is yet another demonstration of precision errors in applied linear algebra.

SVD for Image Denoising

Exercise 15-13.

Let's see if we can extend the concept of low-rank approximation to denoise the Stravinsky picture. The goal of this exercise is to add noise and inspect the SVD results, and then the following exercise will involve "projecting out" the corruption.

The noise here will be a spatial sine wave. You can see the noise and corrupted image in Figure 15-12.

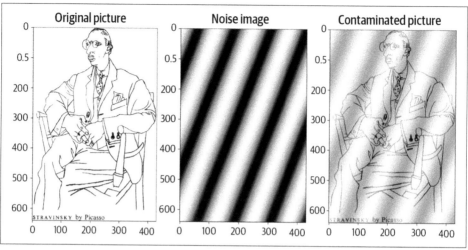

Figure 15-12. Preparation for Exercise 15-13

I will now describe how to create a 2D sine wave (also called a *sine grating*). This is a good opportunity to practice your math-to-code translation skills. The formula for a 2D sine grating is:

$$\mathbf{Z} = \sin\left(2\pi f(\mathbf{X}\cos\left(\theta\right) + \mathbf{Y}\sin\left(\theta\right))\right)$$

In this formula, f is the frequency of the sine wave, θ is a rotation parameter, and π is the constant 3.14.... $\mathbf{X}$ and $\mathbf{Y}$ are grid locations on which the function is evaluated, which I set to be numbers from -100 to 100 with the number of steps set to match the size of the Stravinsky picture. I set $f = .02$ and $\theta = \pi/6$.

Before moving on to the rest of the exercise, I encourage you to spend some time with the sine grating code by exploring the effects of changing the parameters on the resulting image. However, please use the parameters I wrote previously to make sure you can reproduce my results that follow.

Next, corrupt the Stravinsky picture by adding the noise to the image. You should first scale the noise to a range of 0 to 1, then add the noise and the original picture together, and then rescale. Scaling an image between 0 and 1 is achieved by applying the following formula:

$$\widetilde{\mathbf{R}} = \frac{\mathbf{R} - \min(\mathbf{R})}{\max(\mathbf{R}) - \min(\mathbf{R})}$$

OK, now you have your noise-corrupted image. Reproduce Figure 15-13, which is the same as Figure 15-6 but using the noisy image.

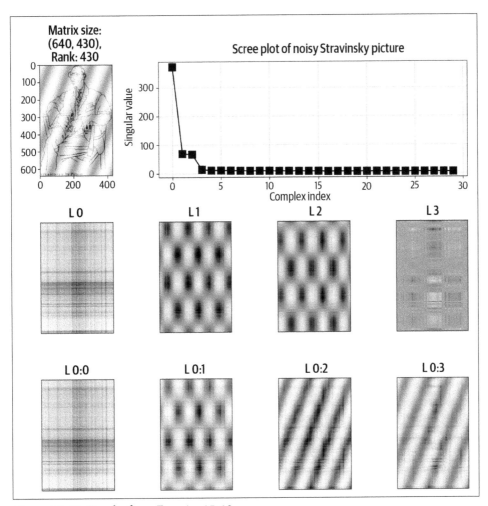

Figure 15-13. Results from Exercise 15-13

Discussion: It's interesting to compare Figure 15-13 with Figure 15-9. Although we created the noise based on one feature (the sine wave grating), the SVD separated the grating into two components of equal importance (roughly equal singular values).[7] Those two components are not sine gratings but instead are vertically oriented patches. Their sum, however, produces the diagonal bands of the grating.

---

7 The likely explanation is that this is a 2D singular *plane*, not a pair of singular *vectors*; any two linearly independent vectors on this plane could be basis vectors, and Python selected an orthogonal pair.

Exercise 15-14.

Now for the denoising. It appears that the noise is contained in the second and third components, so your goal now is to reconstruct the image using all components except for those two. Produce a figure like Figure 15-14.

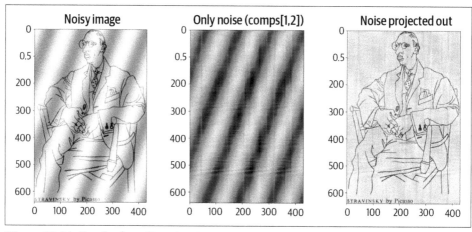

Figure 15-14. Results from Exercise 15-14

Discussion: The denoising is decent but certainly not perfect. One of the reasons for the imperfection is that the noise is not entirely contained in two dimensions (notice that the middle panel of Figure 15-14 does not perfectly match the noise image). Furthermore, the noise projection (the image made from components 1 and 2) has negative values and is distributed around zero, even though the sine grating had no negative values. (You can confirm this by plotting a histogram of the noise image, which I show in the online code.) The rest of the image needs to have fluctuations in values to account for this so that the full reconstruction has only positive values.

## CHAPTER 16

Python Tutorial

As I explained in Chapter 1, this chapter is a crash course on Python programming. It is designed to get you up to speed quickly to follow along with the code in the rest of the book, but it is not designed to be a complete source for Python mastery. If you're looking for a dedicated Python book, then I recommend *Learning Python* by Mark Lutz (O'Reilly).

While working through this chapter, please have a Python session open. (I'll explain how to do that later.) You are not going to learn Python just by *reading* the chapter; you need to read, type into Python, change and test the code, and so on.

Also, in this chapter you should manually key in all the code you see printed here. The code for all other chapters in this book is available online, but I want you to type in the code for this chapter manually. Once you are more familiar with Python coding, then manually typing in lots of code is a tedious waste of time. But when you are first learning to code, you need to *code*—that is, use your fingers to type in everything. Don't just look at code on a page.

Why Python, and What Are the Alternatives?

Python is designed to be a general-purpose programming language. You can use Python to do text analysis, process web forms, create algorithms, and myriad other applications. Python is also widely used in data science and machine learning; for these applications, Python is basically just a calculator. Well, it's an extremely powerful and versatile calculator, but we use Python because we (humans) are not quite smart enough to do all of the numerical calculations in our heads or with pen and paper.

Python is currently (in the year 2022) the most commonly used numerical processing program for data science (other contenders include R, MATLAB, Julia, JavaScript,

SQL, and C). Will Python remain the dominant language for data science? I have no idea, but I doubt it. The history of computer science is ripe with "final languages" that purportedly would last forever. (Have you ever programmed in FORTRAN, COBOL, IDL, Pascal, etc?) But Python is very popular *now*, and you are learning applied linear algebra *now*. Anyway, the good news is that programming languages have strong transfer learning, which means that building Python proficiency will help you learn other languages. In other words, time spent learning Python is invested, not wasted.

IDEs (Interactive Development Environments)

Python is the programming language, and you can run Python in many different applications, which are called *environments*. Different environments are created by different developers with different preferences and needs. Some common IDEs that you might encounter include Visual Studio, Spyder, PyCharm, and Eclipse. Perhaps the most commonly used IDE for learning Python is called Jupyter notebooks.

I wrote the code for this book using Google's Colab Jupyter environment (more on this in the next section). Once you gain familiarity with Python via Jupyter, you can spend some time trying out other IDEs to see if they better suit your needs and preferences. However, I recommend using Jupyter here, because it will help you follow along and reproduce the figures.

Using Python Locally and Online

Because Python is free and lightweight, it can be run on a variety of operating systems, either on your computer or in a cloud server:

Running Python locally
> You can install Python on any major operating system (Windows, Mac, Linux). If you are comfortable with installing programs and packages, then you can install libraries as you need them. For this book, you will mostly need NumPy, matplotlib, SciPy, and sympy.

> But if you are reading this, then your Python skills are probably limited. In that case, I recommend installing Python via the Anaconda software package (*https://www.anaconda.com*). It is free and easy to install, and Anaconda will automatically install all of the libraries you'll need for this book.

Running Python online
> While going through this book, I recommend running Python on the web. Advantages of cloud-located Python include that you don't need to install any-thing locally, you don't need to use your own computing resources, and you can access your code from any browser on any computer and any operating system. I have grown to prefer Google's Colaboratory environment because it

syncs with my Google Drive. That allows me to keep my Python code files on my Google Drive and then open them from *https://colab.research.google.com*. There are several other cloud-based Python environments that you can use if you are avoiding Google services (although I'm not sure that's even possible).

Google Colab is free to use. You'll need a Google account to access it, but that's also free. Then you can simply upload the code files to your Google Drive and open them in Colab.

Working with Code Files in Google Colab

I will now explain how to download and access the Python notebook files for this book. As I wrote earlier, there are no code files for this chapter.

There are two ways to get the book code onto your Google Drive:

- Go to *https://github.com/mikexcohen/LinAlg4DataScience*, click on the green button that says "Code," and then click on "Download ZIP" (Figure 16-1). This will download the code repository, and you can then upload those files onto your Google Drive. Now, from your Google Drive, you can double-click on a file, or you can right-click and select "Open With" and then "Google Colaboratory."

- Go directly to *https://colab.research.google.com*, select the "GitHub" tab, and search for "mikexcohen" in the search bar. You'll find all of my public GitHub repositories; you want the one called "LinAlg4DataScience." From there, you can click on one of the files to open the notebook.

  Note that this is a read-only copy of this notebook; any changes you make will not be saved. Therefore, I recommend copying that file into your Google Drive.

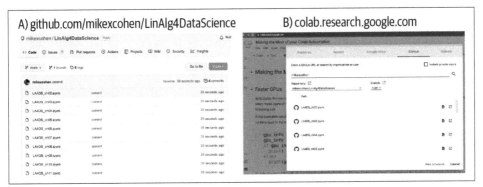

Figure 16-1. Getting the code from GitHub (left) into Colab (right)

Now that you know how to import the book's code files into Google Colab, it's time to start with a fresh notebook to begin working on this chapter. Click on the menu option "File" and then "New Notebook" to create a new notebook. It will be called

"Untitled1.ipynb" or something similar. (The extension *ipynb* stands for "interactive Python notebook.") I recommend changing the name of the file by clicking on the filename at the top-left of the screen. By default, new files are placed in your Google Drive in the "Colab Notebooks" folder.

Variables

You can use Python as a calculator. Let's try it; type the following into a code cell:

```
4 + 5.6
```

Nothing happens when you type that code into a cell. You need to tell Python to run that code. You do that by pressing Ctrl-Enter (Command-Enter on a Mac) on your keyboard while that cell is *active* (a code cell is active if you see the cursor blinking inside the cell). There are also menu options for running the code in a cell, but coding is easier and faster when using keyboard shortcuts.

Take a moment to explore the arithmetic. You can use different numbers, parentheses for grouping, and different operations like -, /, and *. Also notice that the spacing does not affect the result: 2*3 is the same thing as 2 * 3. (Spacing is important for other aspects of Python coding; we'll get to that later.)

Working with individual numbers is not scalable for applications. That's why we have *variables*. A variable is a name that refers to data stored in memory. It's analogous to how languages use words to refer to objects in the real world. For example, my name is Mike, but I am not *Mike*; I am a human being composed of trillions of cells that somehow are capable of walking, talking, eating, dreaming, telling bad jokes, and myriad other things. But that's way too complicated to explain, so for convenience people call me "Mike X Cohen." So, a variable in Python is simply a convenient reference to stored data such as a number, an image, a database, etc.

We create variables in Python by assigning a value to them. Type in the following:

```
var1 = 10
var2 = 20.4
var3 = 'hello, my name is Mike'
```

Running that cell will create the variables. Now you can start to use them! For example, in a new cell, run the following code:

```
var1 + var2

>> 30.4
```

Outputs

The >> you see in code blocks is the result of running the code cell. The text thereafter is what you see on your screen when you evaluate the code in a cell.

Now try this:

```
var1 + var3
```

Ahh, you just got your first Python error! Welcome to the club :) Don't worry, coding errors are extremely common. In fact, *the difference between a good coder and a bad coder is that good coders learn from their mistakes whereas bad coders think that good coders never make mistakes.*

Errors in Python can be difficult to make sense of. Below is the error message on my screen:

```
TypeError                         Traceback (most recent call last)
<ipython-input-3-79613d4a2a16> in <module>()
      3 var3 = 'hello, my name is Mike'
      4
----> 5 var1 + var3

TypeError: unsupported operand type(s) for +: 'int' and 'str'
```

Python indicates the offending line with an arrow. The error message, which will hopefully help us understand what went wrong and how to fix it, is printed at the bottom. In this case, the error message is a TypeError. What does that mean, and what are "types"?

Data Types

It turns out that variables have *types* that describe the kind of data those variables store. Having different types makes computing more efficient, because operations work differently on different kinds of data.

There are many data types in Python. I will introduce four here, and you'll learn more data types as you progress through the book:

Integers
These are called int and are whole numbers, like −3, 0, 10, and 1,234.

Floating-point numbers
These are called float, but that's just a fancy term for a number with a decimal point, like −3.1, 0.12345, and 12.34. Be mindful that floats and ints may look the same to us humans, but they are treated differently by Python functions. For example, 3 is an int, whereas 3.0 is a float.

Strings

These are called `str` and are text. Here, too, be mindful of the distinction between *5* (a string corresponding to the *character* 5) and 5 (an `int` corresponding to the *number* 5).

Lists

A list is a collection of items, each of which can have a different data type.

Lists are quite handy, and are ubiquitous in Python programming. The following code illustrates three important features of lists: (1) they are indicated using square brackets [ ], (2) commas separate list items, and (3) individual list items can have different data types:

```
list1 = [ 1,2,3 ]
list2 = [ 'hello',123.2,[3,'qwerty'] ]
```

The second list shows that lists may contain other lists. In other words, the third element of `list2` is itself a list.

What if you want to access only the second element of `list2`? Individual list elements are extracted using *indexing*, which I will teach you about in the next section.

You can determine the data type using the function `type`. For example, evaluate the following in a new cell:

```
type(var1)
```

Hey, wait, what is a "function"? You can look forward to learning about using and creating functions in the next-next section; first I want to return to the topic of indexing.

What to Call My Variables?

There are a few hard rules about naming variables. Variable names cannot start with numbers (though they may contain numbers), nor can they include spaces or nonalphanumeric characters like ! @#$%^&*(). Underscores _ are allowed.

There are also guidelines for naming variables. The most important guideline is to make variable names meaningful and interpretable. For example, `rawDataMatrix` is a much better variable name than `q`. You might end up creating dozens of variables in your code, and you want to be able to infer the data referenced by a variable from the variable name.

Indexing

Indexing means accessing a specific element in a list (and related data types, including vectors and matrices). Here's how you extract the second element of a list:

```
aList = [ 19,3,4 ]
aList[1]
```

```
>> 3
```

Notice that indexing is done using square brackets after the variable name, and then the number you want to index.

But wait a minute—I wrote that we want the *second* element; why does the code access element 1? That's not a typo! Python is a 0-based indexing language, which means that index 0 is the first element (in this case, the number 19), index 1 is the second element, and so on.

If you are new to 0-based coding languages, then this will seem weird and confusing. I completely sympathize. I wish I could write that it will become second nature after some practice, but the truth is that 0-based indexing will always be a source of confusion and errors. It's just something you have to be mindful of.

How do you access the number 4 in aList? You could index it directly as aList[2]. But Python indexing has a neat feature whereby you can index list elements *backward*. To access the last element of a list, you type aList[-1]. You can think of -1 as wrapping around to the end of the list. Likewise, the penultimate list element is aList[-2], and so on.

Functions

A *function* is a collection of code that you can run without having to type in all the individual pieces of code over and over again. Some functions are short and comprise only a few lines of code, while others have hundreds or thousands of lines of code.

Functions are indicated in Python using parentheses immediately after the function name. Here are a few common functions:

```
type()  # returns the data type
print() # prints text information to the notebook
sum()   # adds numbers together
```

Comments

Comments are pieces of code that are ignored by Python. Comments help you and others interpret and understand the code. Comments in Python are indicated using the # symbol. Any text after the # is ignored. You can have comments in their own lines or to the right of a piece of code.

Functions may take inputs and may provide outputs. The general anatomy of a Python function looks like this:

```
output1,output2 = functionname(input1,input2,input3)
```

Going back to the previous functions:

```
dtype = type(var1)
print(var1+var2)
total = sum([1,3,5,4])

>> 30.4
```

`print()` is a very useful function. Python only prints output of the final line in a cell, and only when that line does not involve a variable assignment. For example, write the following code:

```
var1+var2
total = var1+var2
print(var1+var2)
newvar = 10

>> 30.4
```

There are four lines of code, so you might have expected Python to give four outputs. But only one output is given, corresponding to the `print()` function. The first two lines do not print their output because they are not the final line, while the final line does not print its output because it is a variable assignment.

Methods as Functions

A *method* is a function that is called directly on a variable. Different data types have different methods, meaning that a method that works on lists may not work on strings.

The list data type, for example, has a method called `append` that adds an extra element to an existing list. Here's an example:

```
aSmallList = [ 'one','more' ]
print(aSmallList)

aSmallList.append( 'time' )
print(aSmallList)

>> ['one','more']
['one','more','time']
```

Note the syntax formatting: methods are similar to functions in that they have parentheses and (for some methods) input arguments. But methods are attached to the variable name with a period, and can directly change the variable without an explicit output.

Take a moment to change the code to use a different data type—for example, a string instead of a list. Rerunning the code will generate the following error message:

```
AttributeError: 'str' object has no attribute 'append'
```

This error message means that the string data type does recognize the `append` function (an *attribute* is a property of a variable; a method is one such attribute).

Methods are a core part of object-oriented programming and classes. These are aspects of Python that would be covered in a dedicated Python book, but don't worry —you don't need a full understanding of object-oriented programming to learn linear algebra with this book.

Writing Your Own Functions

There are *many* functions available in Python. Too many to count. But there will never be that perfect function that does exactly what you need. So you will end up writing your own functions.

Creating your own functions is easy and convenient; you use the built-in keyword `def` to define the function (a *keyword* is a reserved name that cannot be redefined as a variable or function), then state the function name and any possible inputs, and end the line with a colon. Any lines thereafter are included in the function *if they are indented with two spaces*.[1] Python is *very* particular about spacing at the beginning of the line (but not particular about spacing elsewhere in the line). Any outputs are indicated by the `return` keyword.

Let's start with a simple example:

```
def add2numbers(n1,n2):
  total = n1+n2
  print(total)
  return total
```

This function takes two inputs and computes, prints, and outputs their sum. Now it's time to call the function:

```
s = add2numbers(4,5)
print(s)

>> 9
9
```

Why did the number 9 appear twice? It was printed once because the `print()` function was called inside the function, and then it was printed a second time when I called `print(s)` after the function. To confirm this, try changing the line after calling

---

1 Some IDEs accept two or four spaces; others accept only four spaces. I think two spaces looks cleaner.

the function to `print(s+1)`. (Modifying code to see the effect on the output is a great way to learn Python; just make sure to undo your changes.)

Notice that the variable name assigned to be the output inside the function (`total`) can be different from the variable name I used when calling the function (`s`).

Writing custom functions allows for a lot of flexibility—for example, setting optional inputs and default parameters, checking inputs for data type and consistency, and so on. But a basic understanding of functions will suffice for this book.

When to Write Functions?

If you have some lines of code that need to be run dozens, hundreds, or maybe billions of times, then writing a function is definitely the way to go. Some people really like writing functions and will write dedicated functions even if they are called only once.

My preference is to create a function only when it will be called multiple times in different contexts or parts of the code. But you are the master of your own code, and as you gain more coding experience, you can choose when to put code into functions.

Libraries

Python is designed to be easy and fast to install and run. But the downside is that the basic install of Python comes with a small number of built-in functions.

Developers therefore create collections of functions focused on a specific topic, which are called *libraries*. Once you import a library into Python, you can access all the functions, variable types, and methods that are available in that library.

According to a Google search, there are over 130,000 libraries for Python. Don't worry, you won't need to memorize all of them! In this book, we will use only a few libraries that are designed for numerical processing and data visualization. The most important library for linear algebra is called *NumPy*, which is a pormanteau of "numerical Python."

Python libraries are separate from the basic Python installation, which means you need to download them from the web and then import them into Python. That makes them available to use inside Python. You need to download them only once, but you need to reimport them into Python in each Python session.[2]

---

[2] If you installed Python via Anaconda or if you are using Google's Colab environment, you won't need to download any libraries for this book, but you will need to import them.

NumPy

To import the NumPy library into Python, type:

```
import numpy as np
```

Notice the general formulation of importing libraries: import libraryname as abbreviation. The abbreviation is a convenient shortcut. To access functions in a library, you write the abbreviated name of the library, a period, and the name of the function. For example:

```
average = np.mean([1,2,3])
sorted1 = np.sort([2,1,3])
theRank = np.linalg.matrix_rank([[1,2],[1,3]])
```

The third line of code shows that libraries can have sublibraries, or *modules*, nested inside them. In this case, NumPy has many functions, and then there is a library inside NumPy called linalg, which contains many more functions specifically related to linear algebra.

NumPy has its own data type called a *NumPy array*. NumPy arrays initially seem similar to lists in that they both store collections of information. But NumPy arrays store only numbers and have attributes that are useful for mathematical coding. The following code shows how to create a NumPy array:

```
vector = np.array([ 9,8,1,2 ])
```

Indexing and Slicing in NumPy

I'd like to return to the discussion of accessing a single element within a variable. You can access one element of a NumPy array using indexing, exactly the same as how you index a list. In the following code block, I'm using the function np.arange to create an array of integers from −4 to 4. That's not a typo in the code—the second input is 5, but the returned values end at 4. Python often uses *exclusive* upper bounds, meaning that Python counts up to—but *not* including—the final number you specify:

```
ary = np.arange(-4,5)
print(ary)
print(ary[5])

>> [-4 -3 -2 -1  0  1  2  3  4]
1
```

That's all well and good, but what if you want to access, for example, the first three elements? Or every second element? It's time to proceed from *indexing* to *slicing*.

Slicing is simple: specify the start and end indices with a colon in between. Just remember that Python ranges have exclusive upper bounds. Thus, to get the first three elements of our array, we slice up to index 3 + 1 = 4, but then we need to account for the 0-based indexing, which means the first three elements have indices 0, 1, and 2, and we slice them using 0:3:

```
ary[0:3]

>> array([-4, -3, -2])
```

You can index every second element using the skip operator:

```
ary[0:5:2]

>> array([-4, -2, 0])
```

The formulation of indexing with skipping is [start:stop:skip]. You can run through the entire array backward by skipping by −1, like this: ary[::-1].

I know, it's a bit confusing. I promise it will get easier with practice.

Visualization

Many concepts in linear algebra—and most other areas of mathematics—are best understood by seeing them on your computer screen.

Most data visualization in Python is handled by the matplotlib library. Some aspects of graphical displays are IDE dependent. However, all the code in this book works as-is in any Jupyter environment (via Google Colab, another cloud server, or a local installation). If you use a different IDE, then you might need to make a few minor adjustments.

Typing matplotlib.pyplot gets really tedious, and so it's common to abbreviate this library as plt. You can see that in the next code block.

Let's begin with plotting dots and lines. See if you can understand how the following code maps onto Figure 16-2:

```
import matplotlib.pyplot as plt
import numpy as np

plt.plot(1,2,'ko') # 1) plot a black circle
plt.plot([0,2],[0,4],'r--') # 2) plot a line
plt.xlim([-4,4]) # 3) set the x-axis limits
plt.ylim([-4,4]) # 4) set the y-axis limits
plt.title('The graph title') # 5) graph title
```

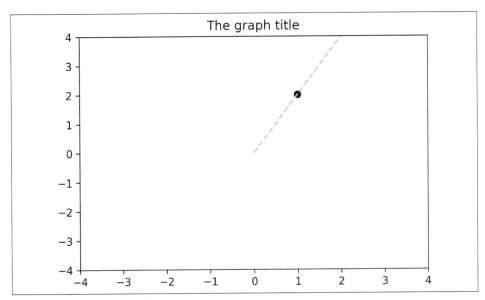

Figure 16-2. Visualizing data, part 1

Did you manage to decode the code? Code line #1 says to plot a black circle (ko—the k is for black and the o is for circle) at XY location 1,2. Code line #2 provides lists of numbers instead of individual numbers. This specifies a line that starts at XY coordinate (0, 0) and ends at coordinate (2, 4). The r-- specifies a red dashed line. Code lines #3 and #4 set the *x*-axis and *y*-axis limits, and, of course, line #5 creates a title.

Before moving forward, take a moment to explore this code. Draw some additional dots and lines, try different markers (hint: explore the letters o, s, and p) and different colors (try r, k, b, y, g, and m).

The next code block introduces subplots and images. A *subplot* is a way of dividing the graphics area (called the *figure*) into a grid of separate axes into which you can draw different visualizations. As with the previous code block, see if you can understand how this code produces Figure 16-3 before reading my description:

```
_,axs = plt.subplots(1,2,figsize=(8,5)) # 1) create subplots
axs[0].plot(np.random.randn(10,5)) # 2) line plot on the left
axs[1].imshow(np.random.randn(10,5)) # 3) image on the right
```

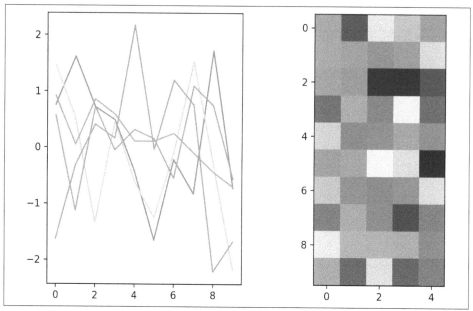

Figure 16-3. Visualizing data, part 2

Code line #1 creates the subplots. The first two inputs into the plt.subplots function specify the grid geometry—in this case, a 1 × 2 matrix of subplots, which means one row and two columns, which means two plots next to each other. The first input specifies the total size of the figure, with the two elements in that tuple corresponding to width and then height in inches. (The sizes are always listed as width,height. A mnemonic for remembering the order is WH for "White House.") The plt.subplots function provides two outputs. The first is a handle to the entire figure, which we don't need, so I used an underscore in lieu of a variable name. The second output is a NumPy array that contains handles to each axis. A *handle* is a special type of variable that points to an object in the figure.

Now for code line #2. This should look familiar to the previous code block; the two new concepts are plotting to the specific axis instead of the entire figure (using plt.) and inputting a matrix instead of individual numbers. Python creates a separate line for each column of the matrix, which is why you see five lines in Figure 16-3.

Finally, code line #3 shows how to create an image. Matrices are often visualized as images, as you learned in Chapter 5. The color of each little block in the image is mapped onto a numerical value in the matrix.

Well, there's *a lot* more that could be said about creating graphics in Python. But I hope this introduction is enough to get you started.

Translating Formulas to Code

Translating mathematical equations into Python code is sometimes straightforward and sometimes difficult. But it is an important skill, and you will improve with practice. Let's start with a simple example in Equation 16-1.

Equation 16-1. An equation

$$y = x^2$$

You might think that the following code would work:

```
y = x**2
```

But you'd get an error message (`NameError: name x is not defined`). The problem is that we are trying to use a variable x before defining it. So how do we define x? In fact, when you look at the mathematical equation, you defined x without really thinking about it: x goes from negative infinity to positive infinity. But you don't draw the function out that far—you would probably choose a limited range to draw that function, perhaps −4 to +4. That range is what we to specify in Python:

```
x = np.arange(-4,5)
y = x**2
```

Figure 16-4 shows the plot of the function, created using `plt.plot(x,y,'s-')`.

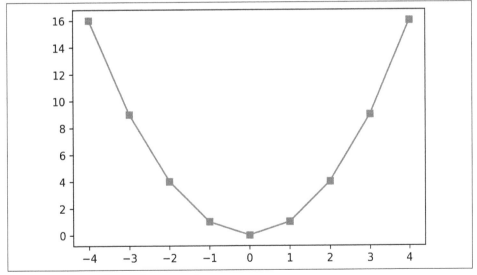

Figure 16-4. Visualizing data, part 3

It looks OK, but I think it's too choppy; I would like the line to look smoother. We can accomplish this by increasing the resolution, which means having more points between −4 and +4. I'll use the function np.linspace(), which takes three inputs: the start value, the stop value, and the number of points in between:

```
x = np.linspace(-4,4,42)
y = x**2
plt.plot(x,y,'s-')
```

Now we have 42 points linearly (evenly) spaced between −4 and +4. That makes the plot smoother (Figure 16-5). Note that np.linspace outputs a vector that ends at +4. This function has inclusive bounds. It is a little confusing to know which functions have inclusive and which have exclusive bounds. Don't worry, you'll get the hang of it.

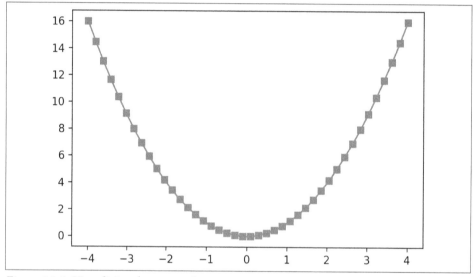

Figure 16-5. Visualizing data, part 4

Let's try another function-to-code translation. I will also use this opportunity to introduce you to a concept called *soft coding*, which means creating variables for parameters that you might want to change later.

Please translate the following mathematical function into code and generate a plot before looking at my code that follows:

$$f(x) = \frac{\alpha}{1 + e^{-\beta x}}$$

$$\alpha = 1.4$$

$$\beta = 2$$

This function is called a *sigmoid* and is used often in applied math, for example as a nonlinear activation function in deep learning models. α and β are parameters of the equation. Here I've set them to specific values. But once you have the code working, you can explore the effects of changing those parameters on the resulting graph. In fact, using code to understand math is, IMHO,[3] the absolute best way to learn mathematics.

There are two ways you can code this function. One is to put the numerical values for α and β directly into the function. This is an example of *hard coding* because the parameter values are directly implemented in the function.

An alternative is to set Python variables to the two parameters, and then use those parameters when creating the mathematical function. This is *soft coding*, and it makes your code easier to read, modify, and debug:

```
x = np.linspace(-4,4,42)
alpha = 1.4
beta  = 2

num = alpha # numerator
den = 1 + np.exp(-beta*x) # denominator
fx  = num / den
plt.plot(x,fx,'s-');
```

Notice that I've split up the function creation into three lines of code that specify the numerator and denominator, and then their ratio. This makes your code cleaner and easier to read. Always strive to make your code easy to read, because it (1) reduces the risk of errors and (2) facilitates debugging.

Figure 16-6 shows the resulting sigmoid. Take a few moments to play with the code: change the x variable limits and resolution, change the `alpha` and `beta` parameter values, maybe even change the function itself. *Mathematics is beautiful, Python is your canvas, and code is your paintbrush!*

---

3 I'm told that this is millenial lingo for "in my humble opinion."

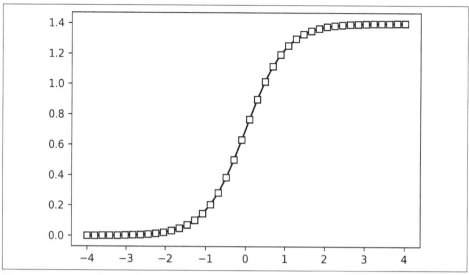

Figure 16-6. Visualizing data, part 5

Print Formatting and F-Strings

You already know how to print out variables using the `print()` function. But that's just for printing out one variable with no other text. F-strings give you more control over the output format. Observe:

```
var1 = 10.54
print(f'The variable value is {var1}, which makes me happy.')

>> The variable value is 10.54, which makes me happy.
```

Note the two key features of the f-string: the initial `f` before the first quote mark and the curly brackets {} encasing variable names that get replaced with variable values.

The next code block further highlights the flexibility of f-strings:

```
theList = ['Mike',7]
print(f'{theList[0]} eats {theList[1]*100} grams of chocolate each day.')

>> Mike eats 700 grams of chocolate each day.
```

Two key points to learn from this example: (1) don't worry, I don't actually eat that much chocolate (well, not every day), and (2) you can use indexing and code inside the curly brackets, and Python will print out the result of the computation.

One final feature of f-string formatting:

```
pi = 22/7
print(f'{pi}, {pi:.3f}')
```

```
>> 3.142857142857143, 3.143
```

The key addition in that code is the `:.3f`, which controls the formatting of the output. This code tells Python to print out three numbers after the decimal point. See what happens when you change the 3 to another integer and what happens when you include an integer before the colon.

There are many other formatting options—and other ways to have flexible text outputs—but the basic implementation of f-strings is all you need to know for this book.

Control Flow

The power and flexibility of programming come from endowing your code with the ability to adapt its behavior depending on the state of certain variables or user inputs. Dynamism in code comes from *control flow* statements.

Comparators

Comparators are special characters that allow you to compare different values. The outcome of a comparator is a data type called a *Boolean*, which takes one of two values: `True` or `False`. Here are a few examples:

```
print( 4<5 ) # 1
print( 4>5 ) # 2
print( 4==5 ) # 3
```

The outputs of these lines are `True` for #1 and `False` for #2 and #3.

That third statement contains a double-equals sign. It's very different from a single equals sign, which you already know is used to assign values to a variable.

Two more comparators are `<=` (less than or equal to) and `>=` (greater than or equal to).

If Statements

If statements are intuitive because you use them all the time: *If I'm tired, then I will rest my eyes.*

The basic `if` statement has three parts: the `if` keyword, the *conditional statement*, and the *code content*. A conditional statement is a piece of code that evaluates to true or false, followed by a colon (:). If the conditional is true, then all the code beneath and indented is run; if the conditional is false, none of the indented code is run, and Python will continue running code that is not indented.

Here is an example:

```
var = 4
if var==4:
  print(f'{var} equals 4!')

print("I'm outside the +for+ loop.")

>> 4 equals 4!
I'm outside the +for+ loop.
```

And here is another example:

```
var = 4
if var==5:
  print(f'{var} equals 5!')

print("I'm outside the +for+ loop.")

>> I'm outside the +for+ loop.
```

The first message is skipped because 4 does not equal 5; therefore, the conditional statement is false, and therefore Python ignores all of the indented code.

elif and else

Those two examples show the basic if statement form. If statements can include additional conditionals to increase the sophistication of the flow of information. Before reading my explanation of the following code and before typing this into Python on your computer, try to understand the code and make a prediction about what messages will print out:

```
var = 4

if var==5:
  print('var is 5') # code 1
elif var>5:
  print('var > 5') # code 2
else:
  print('var < 5') # code 3

print('Outside the if-elif-else')
```

When Python encounters a code statement like this, it proceeds from top to bottom. So, Python will start with the first conditional after the if. If that conditional is true, Python will run code 1 and then *skip all following conditionals*. That is, as soon as Python encounters a true conditional, the indented code is run and the if statement ends. It doesn't matter if subsequent conditionals are also true; Python won't check them or run their indented code.

If the first conditional is false, Python will proceed to the next conditional, which is elif (short for "else if"). Again, Python will run the subsequent indented code if the conditional is true, or it will skip the indented code if the conditional is false. This code example shows one elif statement, but you can have multiple such statements.

The final else statement has no conditional. This is like the "plan B" of the if statement: it is run if all the previous conditionals are false. If at least one of the conditionals is true, then the else code is not evaluated.

The output of this code example is:

```
var <5
Outside the if-elif-else
```

Multiple conditions

You can combine conditionals using and and or. It's the coding analog to "If it rains *and* I need to walk, then I'll bring an umbrella." Here are a few examples:

```
if 4==4 and 4<10:
  print('Code example 1.')

if 4==5 and 4<10:
  print('Code example 2.')

if 4==5 or 4<10:
  print('Code example 3.')


>> Code example 1.
Code example 3.
```

The text Code example 2 did not print because 4 does not equal 5. However, when using or, then *at least one* of the conditionals is true, so the subsequent code was run.

For Loops

Your Python skills are now sufficient to print out the numbers 1–10. You could use the following code:

```
print(1)
print(2)
print(3)
```

And so on. But that is not a scalable strategy—what if I asked you to print out numbers up to a million?

Repeating code in Python is done through *loops*. The most important kind of loop is called a *for loop*. To create a for loop, you specify an iterable (an *iterable* is a variable used to iterate over each element in that variable; a list can be used as an iterable) and then any number of lines of code that should be run inside the for loop. I'll start with a very simple example, and then we'll build on that:

```
for i in range(0,10):
    print(i+1)
```

Running that code will output the numbers 0 through 10. The function range() creates an iterable object with its own data type called *range*, which is often used in for loops. The range variable contains integers from 0 through 9. (Exclusive upper bound! Also, if you start counting at 0, then you don't need the first input, so range(10) is the same as range(0,10)). But my instructions were to print the numbers 1 through 10, so we need to add 1 inside the print function. This example also highlights that you can use the iteration variable as a regular numerical variable.

for loops can iterate over other data types. Consider the following example:

```
theList = [ 2,'hello',np.linspace(0,1,14) ]
for item in theList:
    print(item)
```

Now we're iterating over a list, and the looping variable item is set to each item in the list at each iteration.

Nested Control Statements

Nesting flow-control statements inside other flow-control statements gives your code an additional layer of flexibility. Try to figure out what the code does and make a prediction for its output. Then type it into Python and test your hypothesis:

```
powers = [0]*10

for i in range(len(powers)):
    if i%2==0 and i>0:
        print(f'{i} is an even number')

    if i>4:
        powers[i] = i**2

print(powers)
```

I haven't yet taught you about the % operator. That's called the *modulus operator*, and it returns the remainder after division. So 7%3 = 1, because 3 goes into 7 twice with a remainder of 1. Likewise, 6%2 = 0 because 2 goes into 6 three times with a remainder of 0. In fact, k%2 = 0 for *all* even numbers and k%2 = 1 for *all* odd numbers. Therefore, a statement like i%2==0 is a way to test whether the numeric variable i is even or odd.

Measuring Computation Time

When writing and evaluating code, you will often want to know how long the computer takes to run certain pieces of code. There are several ways to measure elapsed time in Python; one simple method is shown here, using the time library:

```
import time

clockStart = time.time()
# some code here...
compTime = time.time() - clockStart
```

The idea is to query the operating system's local time twice (this is the output of the function `time.time()`): once before running some code or functions, and once after running the code. The difference in clock times is the computation time. The result is the elapsed time in seconds. It's often handy to multiply the result by 1,000 to print the results in milliseconds (ms).

Getting Help and Learning More

I'm sure you've heard the phrase "Math is not a spectator sport." Same goes for coding: the only way to learn to code is by coding. You will make lots of mistakes, get frustrated because you can't figure out how to get Python to do what you want, see lots of errors and warning messages that you can't decipher, and just get really irritated at the universe and everything in it. (Yeah, you know the feeling I'm referring to.)

What to Do When Things Go Awry

Please allow me the indulgence to tell a joke: four engineers get into a car, but the car won't start. The mechanical engineer says, "It's probably a problem with the timing belt." The chemical engineer says, "No, I think the problem is the gas/air mixture." The electrical engineer says, "It sounds to me like the spark plugs are faulty." Finally, the software engineer says, "Let's just get out of the car and get back in again."

The moral of the story is that when you encounter some unexplainable issues in your code, you can try restarting the *kernel*, which is the engine running Python. That won't fix coding errors, but it may resolve errors due to variables being overwritten or renamed, memory overloads, or system failures. In Jupyter notebooks, you can restart the kernel through the menu options. Be aware that restarting the kernel clears out all of the variables and environment settings. You may need to rerun the code from the beginning.

If the error persists, then search the internet for the error message, the name of the function you're using, or a brief description of the problem you're trying to solve. Python has a huge international community, and there is a multitude of online forums where people discuss and resolve Python coding issues and confusions.

Summary

Mastering a programming language like Python takes years of dedicated study and practice. Even achieving a good beginner's level takes weeks to months. I hope this chapter provided you with enough skills to complete this book. But as I wrote in Chapter 1, if you find that you understand the math but struggle with the code, then you might want to put this book down, get some more Python training, and then come back.

On the other hand, you should also see this book as a way to improve your Python coding skills. So if you don't understand some code in the book, then learning linear algebra is the perfect excuse to learn more Python!

Index

vector-scalar version, 244

finding, 217-220

from eigendecomposition of singular matrix, 227

noise reduction with, 216

real-valued, 226

signs of, determining sign of quadratic form, 230

eigenvectors, 214-215, 234

finding, 220-222

sign and scale indeterminacy of eigenvectors, 221

from eigendecomposition of singular matrix, 227

in null space of the matrix shifted by its eigenvalue, 218

orthogonal, 224

real-valued in symmetric matrices, 226

stored in matrix columns, not rows, 220

element-wise multiplication, 67

(see also Hadamard multiplication)

elif and else statements (Python), 298

equations, systems of, 159-163, 165, 171

converting equations into matrices, 160

solved by matrix inverse, 167

solving with Gauss-Jordan elimination, 167

working with matrix equations, 161

equations, translating to code in Python, 293-296

errors

in Python, 283

squared errors between predicted data and observed data, 181

Euclidean distance, 55

F

f-strings, 296

feature detection, time series filtering and, 52

filtering, 52

image, 120-123

float type, 283

for loops (Python), 299

formulas, translating to code in Python, 293-296

Frobenius norm, 200

full inverse, 130

full or complete QR decomposition, 151

full-rank matrices

no zero-valued eigenvalues, 228

square, computing inverse of, 134

functions (Python), 285-290

libraries of, 288

importing NumPy, 289

indexing and slicing in NumPy, 289

methods as functions, 286

writing your own, 287

G

Gauss-Jordan elimination, 166, 187

matrix inverse via, 167

Gaussian elimination, 165

Gaussian kernel (2D), 122

general linear models, 175-178, 188

setting up, 176-178

in simple example, 183-186

solving, 178-183

exactness of the solution, 179

geometric perspective on least squares, 180

why least squares works, 181

solving with multicollinearity, 199

terminology, 176

generalized eigendecomposition, 232

on covariance matrices in LDA, 261

geometric interpretation of matrix inverse, 141

geometric length (vectors), 17

geometric transforms, 71

using matrix-vector multiplication, 116-120

geometry of vectors, 10

addition and subtraction of vectors, 12

dot product, 21

eigenvectors, 214

vector-scalar multiplication, 15

GLMs (see general linear models)

Google Colab environment, 5, 280

working with code files in, 281

Gram-Schmidt procedure, 149

grid matrix, 135

grid search to find model parameters, 204-205

GS or G-S (see Gram-Schmidt procedure)

H

Hadamard multiplication, 23

of matrices, 67

hard coding, 295

Hermitian, 244

Hilbert matrix, its inverse, and their product, 140

About the Author

Mike X Cohen is an associate professor of neuroscience (*https://oreil.ly/Ee23F*) at the Donders Institute (Radboud University Medical Centre) in the Netherlands. He has over 20 years of experience teaching scientific coding, data analysis, statistics, and related topics, and has authored several online courses (*https://oreil.ly/BurUH*) and textbooks. He has a suspiciously dry sense of humor and enjoys anything purple.

Colophon

The animal on the cover of *Practical Linear Algebra for Data Science* is a nyala antelope, also known as the lowland nyala or simply nyala (*Tragelaphus angasii*). Female and juvenile nyalas are typically a light reddish-brown, while adult males have a dark brown or even grayish coat. Both males and females have white stripes along the body and white spots on the flank. Males have spiral-shaped horns that can grow up to 33 inches long, and their coats are much shaggier, with a long fringe hanging from their throats to their hindquarters and a mane of thick black hair along their spines. Females weigh about 130 pounds, while males can weigh as much as 275 pounds.

Nyalas are native to the woodlands of southeastern Africa, with a range that includes Malawi, Mozambique, South Africa, Eswatini, Zambia, and Zimbabwe. They are shy creatures, preferring to graze in the early morning, late afternoon, or nighttime, and spending most of the hot part of the day resting among cover. Nyalas form loose herds of up to ten animals, though older males are solitary. They are not territorial, though males will fight over dominance during mating.

Nyalas are considered a species of least concern, though cattle grazing, agriculture, and habitat loss pose a threat to them. Many of the animals on O'Reilly covers are endangered; all of them are important to the world.

The cover illustration is by Karen Montgomery, based on an antique line engraving from *Histoire Naturelle*. The cover fonts are Gilroy Semibold and Guardian Sans. The text font is Adobe Minion Pro; the heading font is Adobe Myriad Condensed; and the code font is Dalton Maag's Ubuntu Mono.

Milton Keynes UK
Ingram Content Group UK Ltd.
UKHW052339190524
442887UK00004B/9